Math Illustrations™ *Manual*

Saltire Software
PO Box 230755
Tigard, OR 97281 0755

This material is based upon work supported by the National Science Foundation under Grant No. 0750028

Table of Contents

Getting Started

Welcome

Math Illustrations is an application that lets you create high quality mathematical diagrams for your mathematical worksheets, test and presentations in a fraction of the time.

Simply sketch your geometry, add symbolic constraints and other annotations. Your diagrams are drawn to scale in a flash!

Math Illustrations is based upon work supported by the National Science Foundation under Grant No. 0750028

Need Help?

There are many ways to get help with *Math Illustrations*.

To get you on your way you'll find the *Quick Start Guide,* in PDF format in your Doc directory.

Your installation comes with *Math Illustrations Manual*, in PDF format suitable for creating a hard copy.

In both the printed manual and the embedded Help system ⓘ you can:

- Use the table of contents to get details on using a tool, an icon or a menu.

- Use the index for help on a particular topic, such as "parametric equations".

Inside the *Math Illustrations* Help system you can:

- Use the search tool to find all topics based on a key word, such as "constraints".

- Browse through help with the up - "Previous page", and down - "Next Page" arrows ⇧ ⇩ . This will step you through the help subtopics in a logical sequence.

- You can click the links displayed in colored text to get more information.

Tool Tips:

When you move the cursor over any icon on the screen, the name of the icon appears briefly below the cursor.

Linked Text

Some words in the program's Help system are highlighted and underlined. When you place the cursor over this text, the cursor becomes the hand symbol. This text indicates a link to more information on the subject. Click the text to jump to the related help page.

The Display and How It's Organized

Many of the menu items in the drop down menu bar at the top of the screen correspond to one of the icons or buttons across the top of the display window or in one of the toolboxes.

Icons across the top of the screen comprise the standard Windows **File, Edit, View** and **Help** commands. The construction and calculation tools are displayed along the side of the drawing window. These toolboxes can be displayed on the left or right panel, top or bottom panel, floating in a separate window, or hidden.

The status bar at the bottom of the window displays the following (from left to right):

<Menu Help> <Current Mode> <Cursor Coordinates> <Angle Mode>

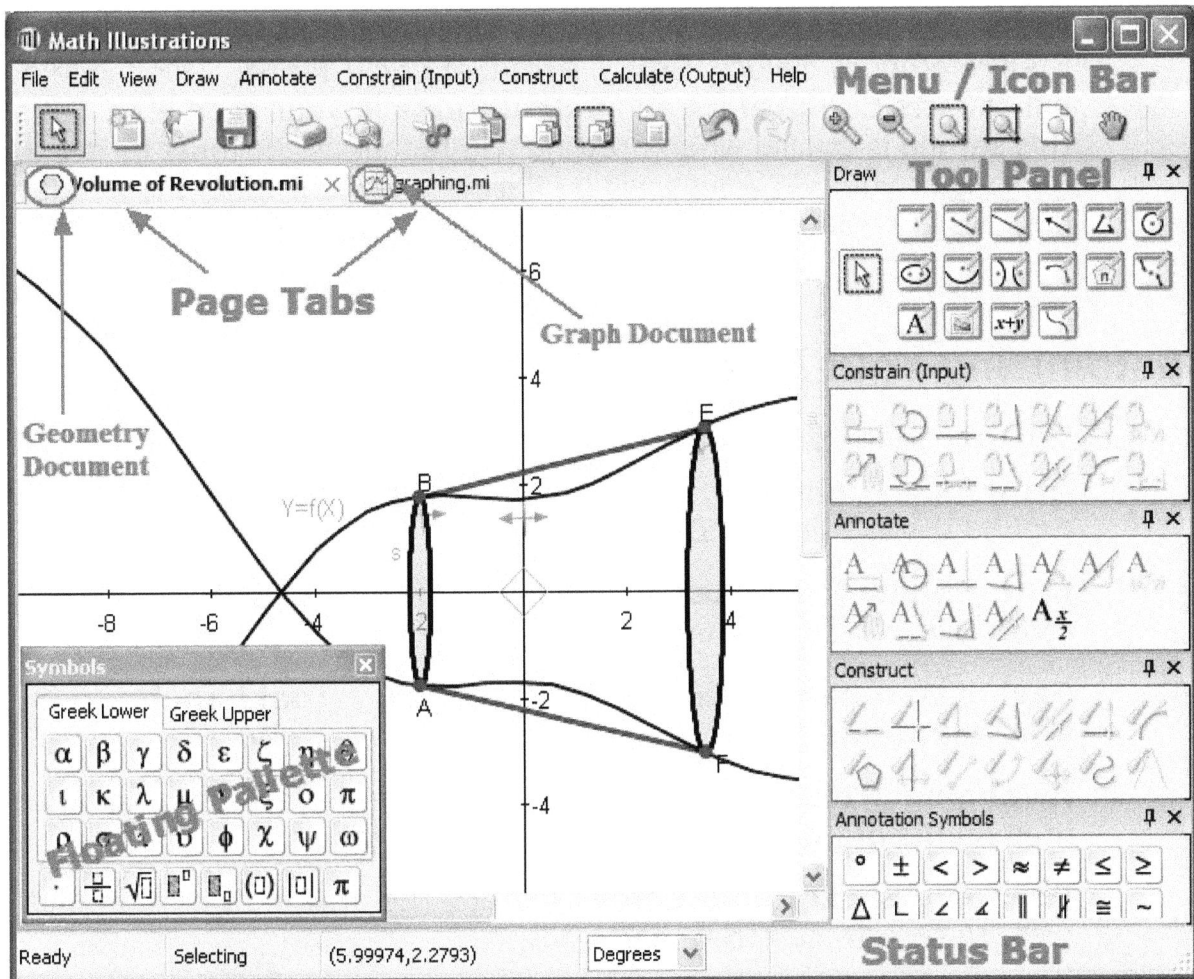

The Status Bar

The status bar at the bottom of the screen prompts the following information (from left to right):

- **Menu Help** - summary of a selected menu item.

- **Current Mode** - Each icon in the drawing toolbox represents a mode. Tools requiring additional inputs after clicking the tool will display further prompts in this field.

- **Cursor Coordinates** - Displays the current coordinates of the cursor in the diagram.

- **Angle Mode** - A drop down window for conveniently changing from

Radians to Degrees and *visa versa*. This default can also be changed in the Preferences dialog - **Edit / Preferences / Math.** This mode not only affects **Constraints** and **Calculations** in your geometry, but also this setting is crucial when drawing trig functions. It must be set to Radians or your sin waves will be flat! You can still label the X axis in degrees or radians from the **Context** menu, **Axis Properties / Units**.

Customizing Your Display

You can arrange the display as it suits you.

Arranging Toolboxes - Anchored or floating toolboxes can be placed around the drawing window.

Hide / Show Toolboxes - You may want to hide toolboxes which you rarely use.

Saving your configuration - Use the **View / Tool Panel Configurations**.

Arranging projects - You can open multiple project files and arrange them in the drawing window using the page tabs.

Changing Background Color - Change the drawing's background color to something other than white.

In the example below six toolboxes are anchored (**Constrain (Input), Construct, Calculate (Output), Draw, Annotate** and **Variables**), one is floating (**Annotation Symbols**), and one (**Symbols**) is hidden. Two slightly different locus examples are displayed for comparison.

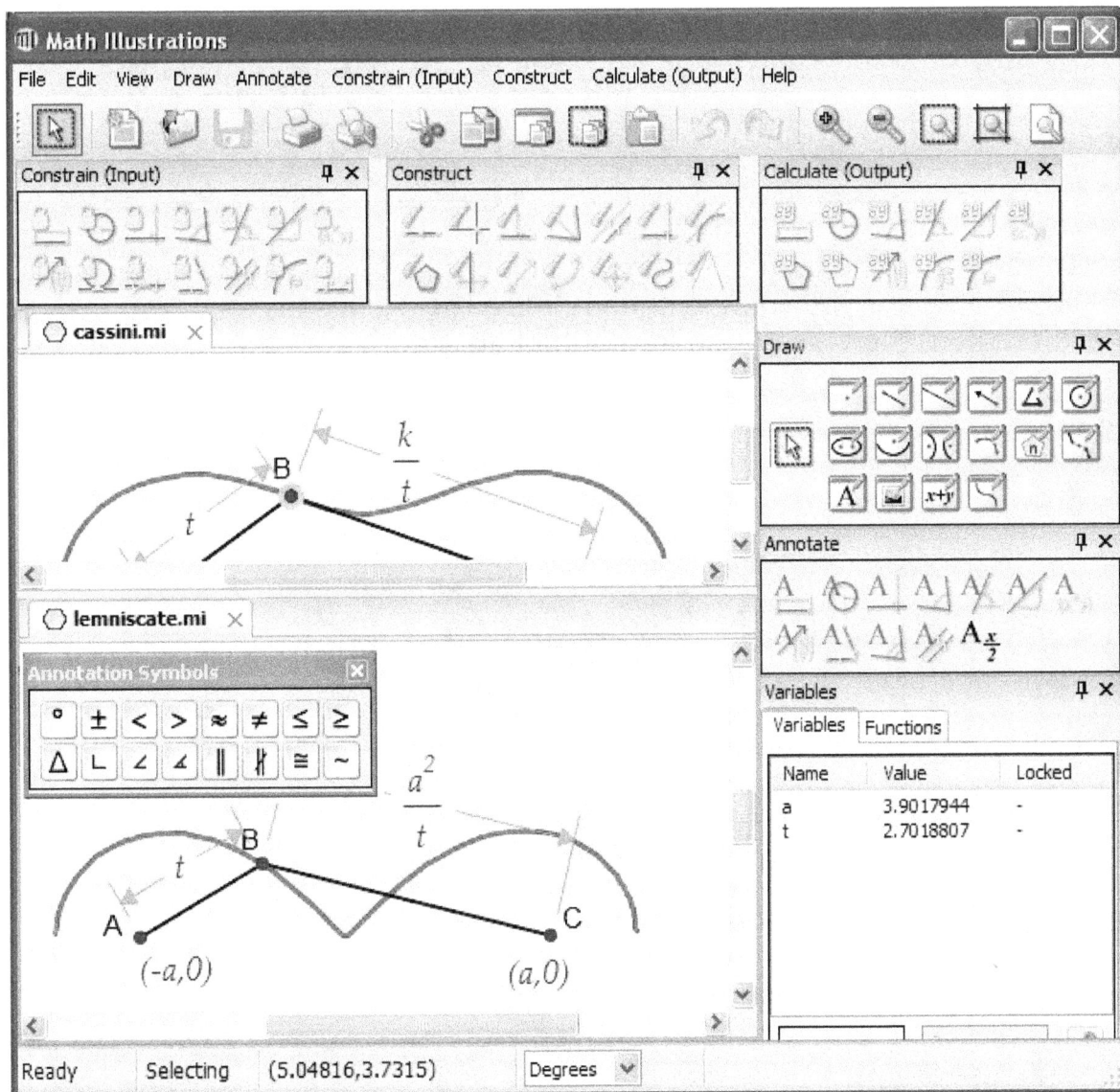

Arranging Toolboxes

You can move the toolboxes around the periphery of the drawing window by clicking the title bar and dragging.

Click the pushpin on the upper right corner of individual toolboxes to make it a "floating" box that you can drag anyplace on the screen. Floating boxes have a colored title bar -

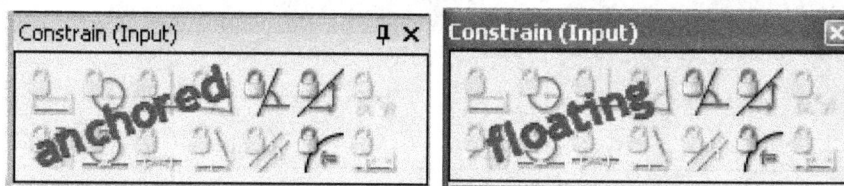

To re-anchor the toolbox, drag the box until a blue shadow appears at the position where you want it, then release the mouse button. You may want to readjust the screen size for optimal viewing.

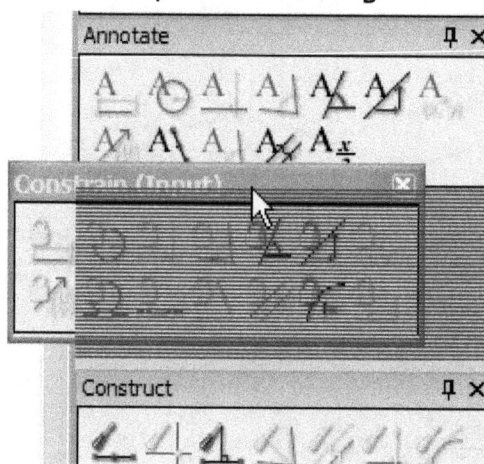

Hiding / Showing Toolboxes

Use the X on the upper right corner of individual toolboxes to hide them.

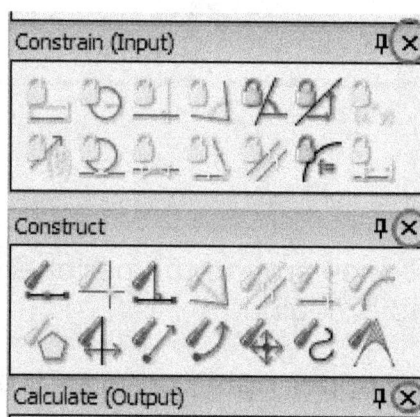

To display a hidden toolbox, select **View / Tool Panels**. The submenu lists the toolboxes and the **Main Toolbar** (the icon strip at the top of the window). Boxes shown are preceded by a check, those without a check are hidden.

Axes	
Grid	
Page Boundaries	Construct
Tool Panels ▶	✔ Main Toolbar
Tool Panel Configurations ▶	✔ Draw
	✔ Constrain (Input)
Output	✔ Annotate
	✔ Construct
Language ▶	✔ Calculate (Output)
	✔ Symbols
	Annotation Symbols
	✔ Variables

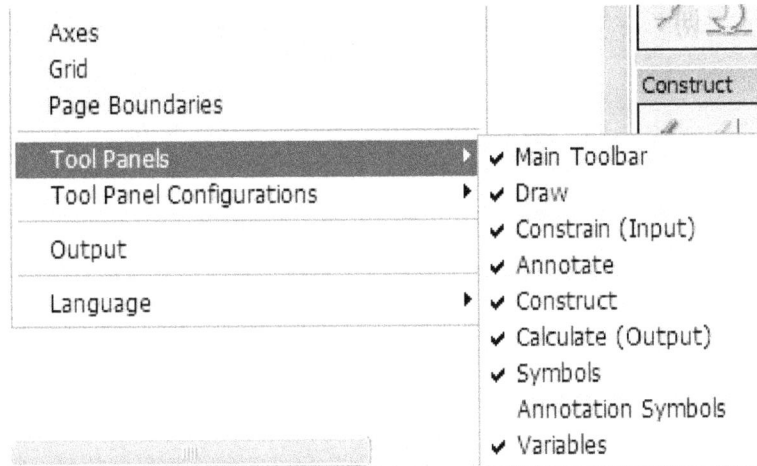

Click a toolbox name to change its state.

Saving the Configuration

After configuring the screen to you preference, you can save this arrangement in case it gets messed up, or perhaps you need the tools arranged differently for different projects. This is easy with the **View / Tool Panel Configurations** menu selection. You can give a name to an arrangement of the toolbars. Several configurations can be saved in a list and referred to as needed.

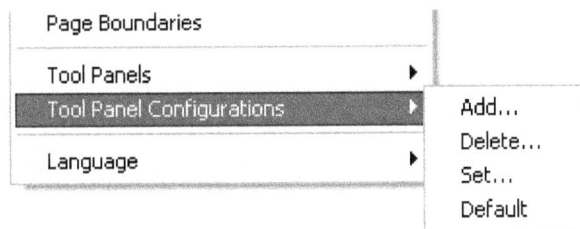

Page Boundaries		
Tool Panels	▶	
Tool Panel Configurations	▶	Add...
		Delete...
Language	▶	Set...
		Default

Add - to save the current screen configuration. You will be prompted for a name to reference this configuration.

Delete - if you no longer will use a certain configuration you can delete it from your list.

Set - to change a configuration which you have saved, simply select the configuration name from your list. Check out the configurations that come with the program.

Set Tool Panel Configuration

Select configuration

Geometry Expressions
Geometry Publishing
My Special Configuration
above&left
left side
right side

OK   Cancel

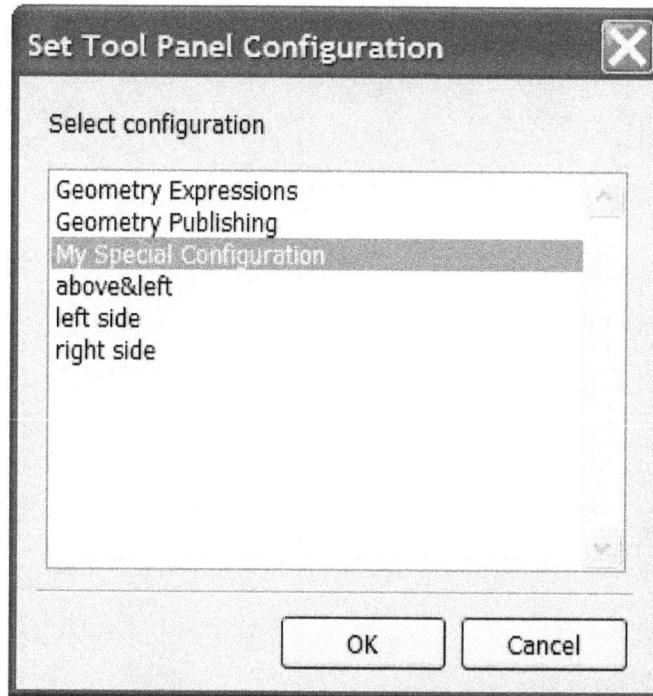

Default - reset tool panels back to the default configuration.

Arranging Project Pages

You can open multiple project files for quick reference. By default files are overlaid. Click a page tab to bring a file to the top.

Comparing drawings side by side - click the page tab and drag it to one side, top, or bottom of the window. A shadow of the drawing gives you an indication of how the drawings will be arranged before you release the mouse button.

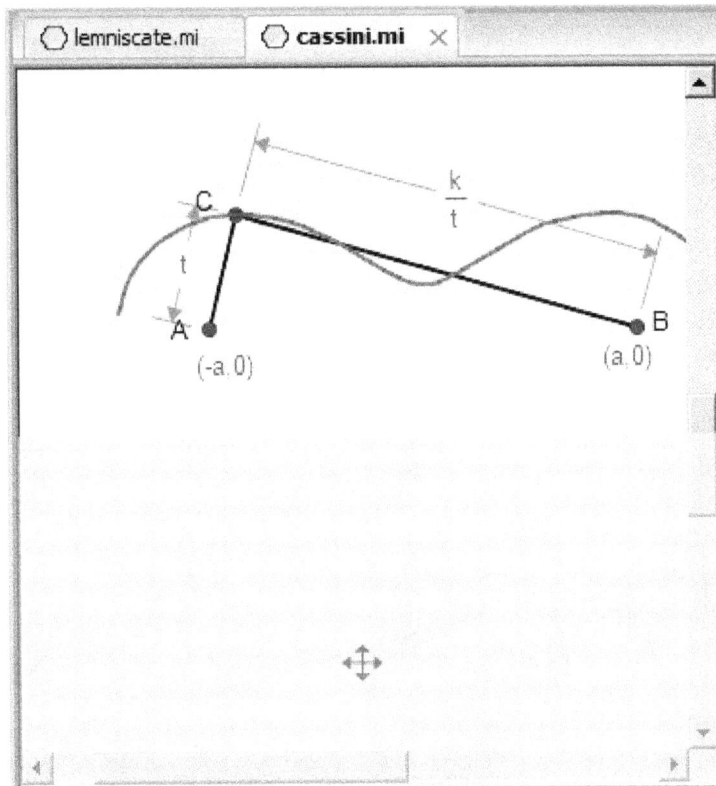

Returning to overlaid configuration - drag one tab to the other title bar. The shadow will appear only on the title bar, then release the mouse button.

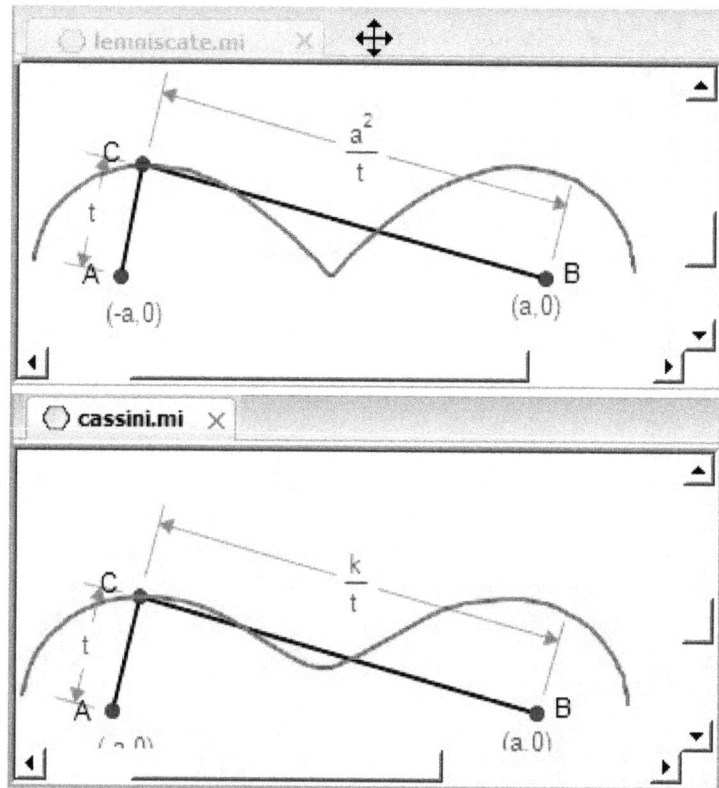

Changing Background Color

You can now change the drawing's background color to something other than white by the following steps:

1. Select **Edit / Preferences...** (or in the Mac version, **Math Illustrations / Preferences...**)

2. Click **Grid, Axis, Page** button

3. Change **Fill Color** in **Background** section. The color will be applied to all pages.

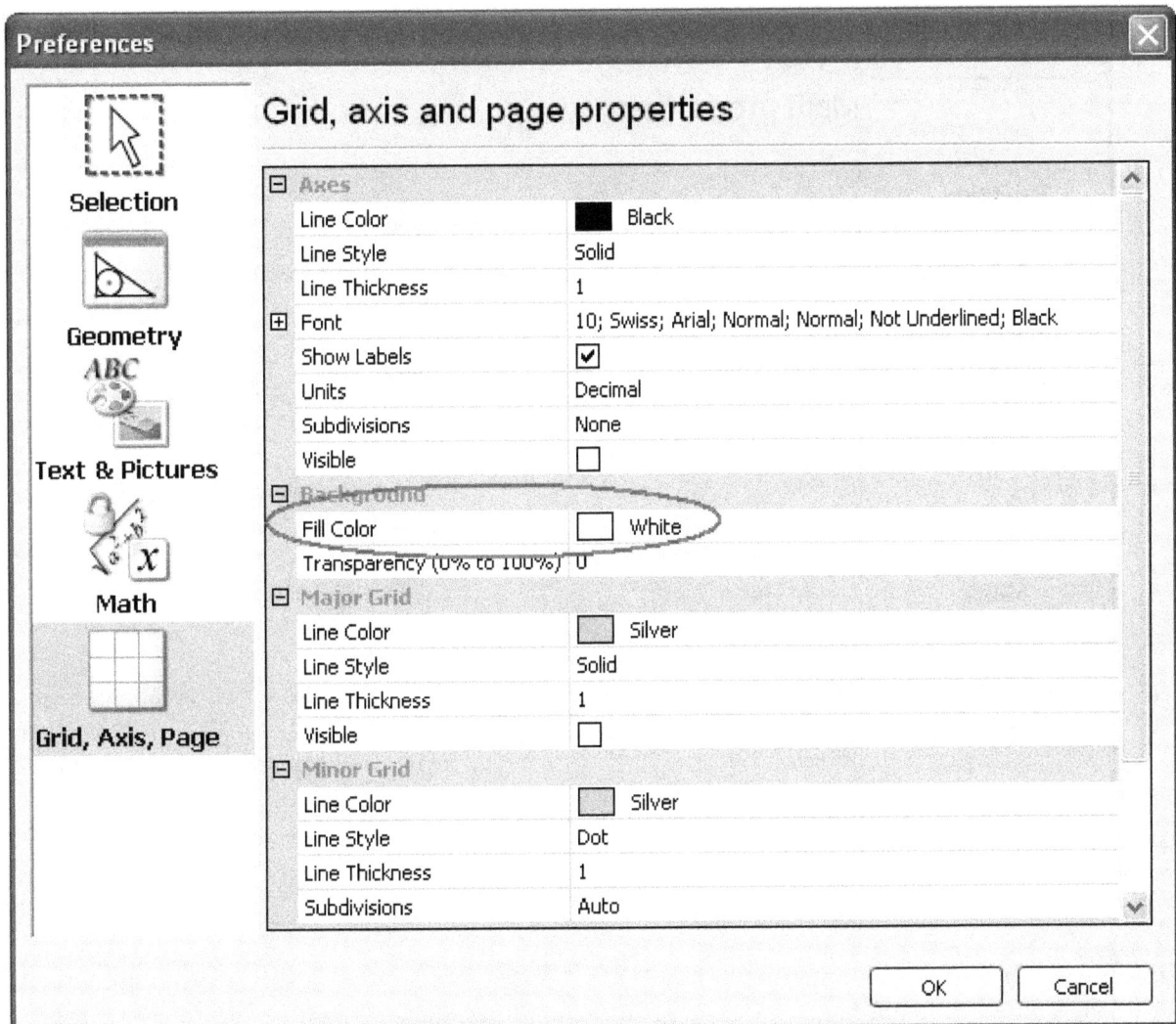

Adjusting the Default Settings

Select **Preferences** from the **Edit** menu (on Mac, it's under *Math Illustrations*) to modify the program's defaults.

The default settings are grouped by type, listed on the left side of the **Preferences** dialog. Click the icon to display the desired page.

These settings are also available for editing a selected object or group of objects individually without changing the defaults, using the Selection Context Menu.

Preferences

Math properties

Annotation	
Line Color	Black
Line Style	Solid
Line Thickness	1
⊞ Font	10; Swiss; Arial; Normal; Normal; Not Underline
Calculation (Output)	
Line Color	(140,140,140)
Line Style	Solid
Line Thickness	1
⊞ Font	10; Swiss; Arial; Normal; Normal; Not Underline
Line Equation Style	ax+by+c=0
Constrain (Input)	
Line Color	(130,130,255)
Line Style	Solid
Line Thickness	1
⊞ Font	10; Swiss; Arial; Normal; Normal; Not Underline
Expression	
⊞ Font	10; Swiss; Arial; Normal; Normal; Not Underline
Pinned	☐
Math	
Angle Mode	Radians
Intermediate Variable Complexity (2 to 100)	15

Selection
Geometry
Text & Pictures
Math
Grid, Axis, Page

OK   Cancel

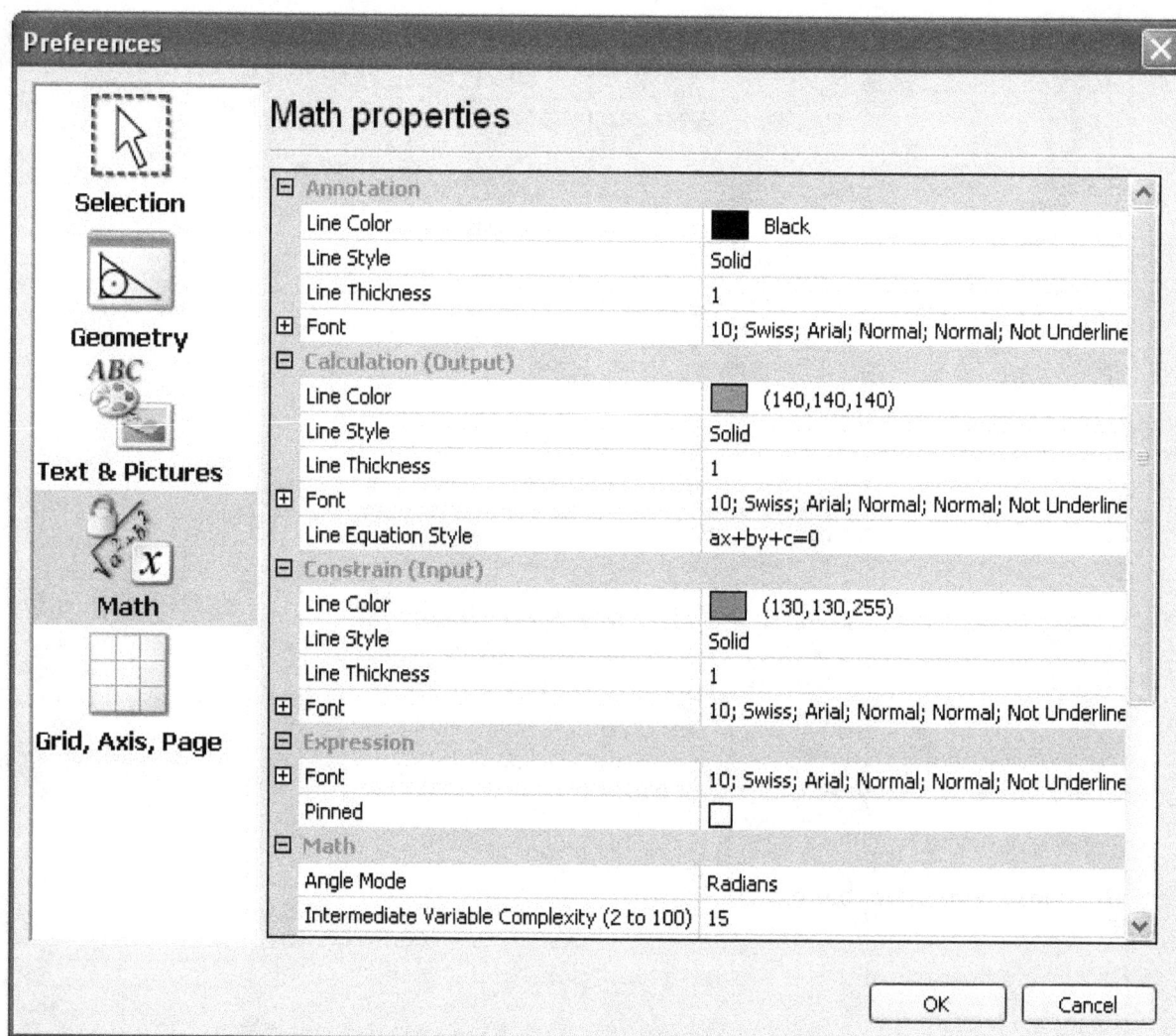

Selection	set the line color and style for each selection type.
Geometry	set font related properties for labels; color and size / style for other geometric elements.
Text & Pictures	set font properties for text, the rotation angle and the transparency level of an inserted picture, and the Pinned state for **Text** and **Pictures**. Pinned **Text** and **Pictures** will not move relative to the **Page Boundaries** as the result of a **Scale** operation.

Math	set the properties for alphanumeric input and output; mathematical calculation defaults.
Grid, Axis, Page	set properties of the Major and Minor Grid, the coordinate Axes, the background color, and the Page Boundary lines.

To see the possible values for each property, click the row. An icon will appear at the right end of the row (except the Point Size selection under the Font property -you can enter the point size directly).

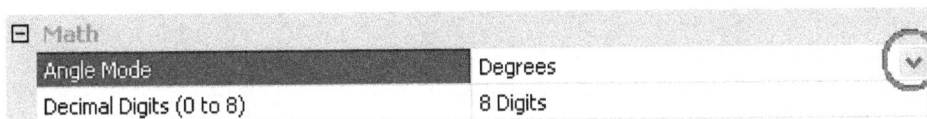

⊟ Math		
Angle Mode	Degrees	⌄
Decimal Digits (0 to 8)	8 Digits	

Click the icon to display the selection dialog [...] or drop-down menu of choices ⌄.

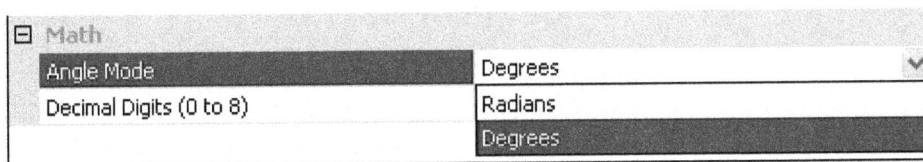

⊟ Math		
Angle Mode	Degrees	⌄
Decimal Digits (0 to 8)	Radians	
	Degrees	

Changing a default for a type of drawing entity will apply to all entities of that type except ones whose properties have been individually set, by selecting it and then choosing **Edit / Properties** or right clicking **All Properties** from the **Selection Context** menu. Also, text, pictures or expressions that were individually pinned ⊙Circle or unpinned ⊶Circle will not be affected by changes to the default Pinned settings.

File Handling

Math Illustrations uses standard Windows file **Open** and **Save** operations. Save your files regularly with the handy 💾 icon at the top of the screen.

The data files generated from your drawings will have the extension ".mi".

You may create multiple data files and have them open in a session. Each file is on separate page with the tabs across the top of the drawing window. Click the tab to view the file.

If you are preparing a multi-paged lesson, you can save the pages together as a **Workbook**, with the file extension ".miw". This is a completely separate file from the .mi files.

The **Open** / **Save** (**As**) / **Close** Workbook file selections apply only to the workbook. 🖫 and the **File** / **Save** commands will NOT save the workbook files. They only affect the individual .mi files.

Files can also be arranged for comparison viewing.

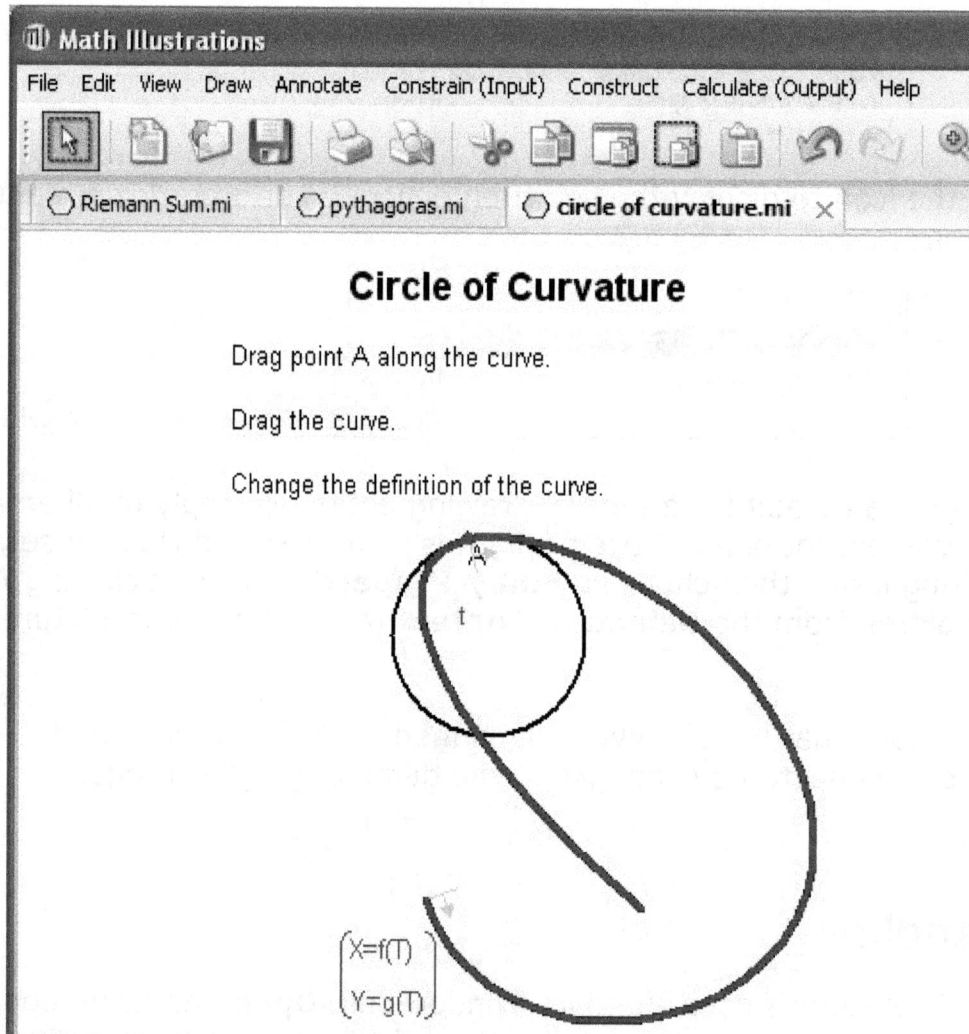

Wookbooks and Individual .mi Files

Workbooks are a handy way of putting lessons together.

- You can make a workbook by opening new tabs (**File / New**).

- You can **Open** .mi files that you have already created to make them part of your workbook.

- You can create pages from a combination of the above.

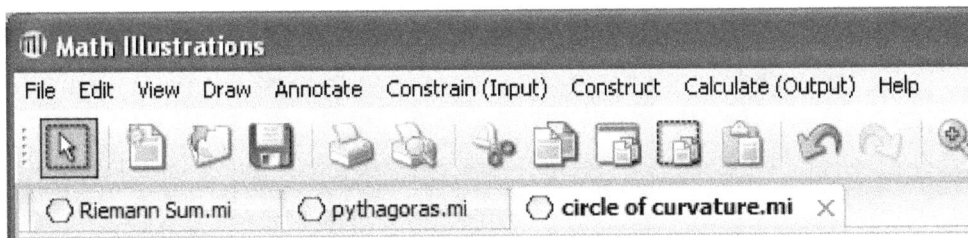

When all of the pages are together on the page tab bar, select **Save Workbook (As)** from the file menu.

! Note: saving the workbook does NOT update any of the individual .mi files displayed in the workbook, since the workbook file is an entirely separate file (.miw). To do this you must use the regular **File / Save** for each page / file if you want to keep the individual .mi file up to date with the workbook page. But, if you forget, you can always save it from the workbook at any time.

Likewise, saving an individual page, **File / Save**, of an open workbook does NOT update the workbook file (.miw). However, if you forget to save the workbook, but save a page as a regular .mi file, you can always open the . mi file again from your open workbook file and resave the workbook.

If you don't want to keep individual copies of all your workbook pages, then you just have to remember to use the **Save Workbook** file selection.

However, to give the workbook pages custom names (instead of unnamed7. mi), you must save the individual page (for example, Lesson 1 Ellipse.mi) at least once. After the first time, you don't need to continue to save the .mi file.

Only one workbook file can be open at a time. If you open a workbook file while other individual files or another workbook file is open, they will be closed, after, of course, prompting you to save them if you have made changes.

Tools

Drawing

Using the Drawing Tools

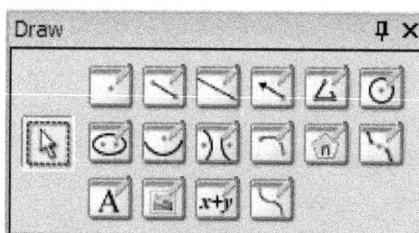

The **Draw** toolbox contains the drawing modes and the Selection Arrow. The drawing modes can also be invoked from the **Draw** menu. Once selected, a mode is active until you change to a different one.

The selection mode is used for invoking all other commands. You can find the active mode by noting which button is pressed or looking at the mode area in the status bar.

Point	Line Segment	Infinite Line	Vector	Polygon	Circle

Ellipse	Parabola	Hyperbola	Arc	N-gon	Curve Approximation

Text	Picture	Expression	Function

Adding a Point

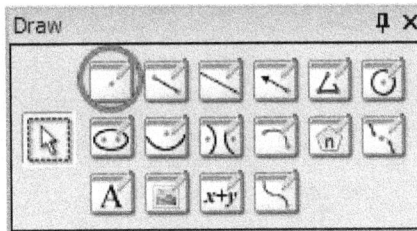

To insert a point in your drawing, follow these steps:

1. Click the **Point** icon ⬚ in the **Drawing** toolbox or select **Point** from the **Draw** menu.

2. Move the crosshairs ┼ into position.

3. Click the mouse to place the point under the crosshair.

When the crosshairs are positioned over some geometry an incidence symbol (bowtie) is displayed around the point and the geometry is highlighted. A click of the mouse will create the point incident to the highlighted geometry.

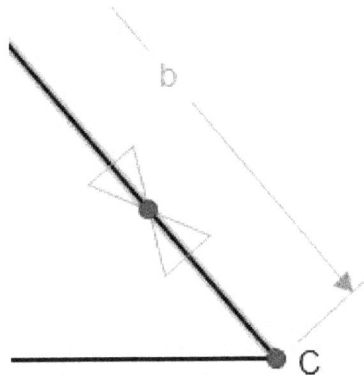

Each point is displayed with a letter label. You can change the label from the Select mode ⬚.

Point is a modal command. You can continue making points until you choose the select arrow or another drawing tool.

Adding Line Segments

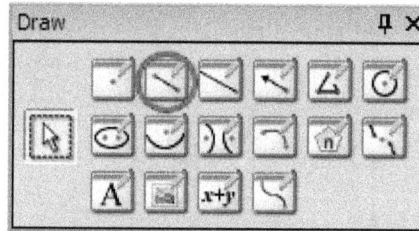

To add line segments to your geometry follow these steps:

1. Click on the **Line Segment** icon in the **Draw** toolbox or select **Line Segment** from the **Draw** menu.

2. Position the cursor in the drawing window.

3. Click the mouse to place each endpoint.

Each line segment is displayed with a letter label for each endpoint.

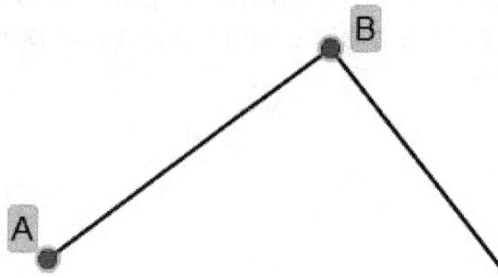

You can change the label from the Select mode .

To abort a line segment in the middle of the drawing operation, hit the "esc" key.

Line Segment is a modal command. You can continue making segments until you choose the select arrow or another drawing tool.

Drawing Lines

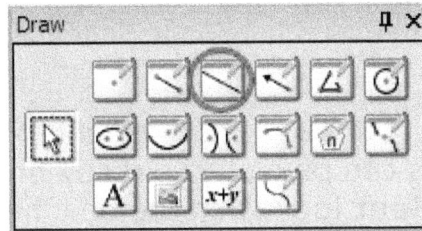

Lines are similar to line segments except they have infinite length.

1. Click the **Infinite Line** icon in the **Draw** toolbox or select **Infinite Line** from the **Draw** menu.

2. Position the line cursor ╪ in the drawing window.

3. Click the cursor to anchor the line at the cursor position. The anchor point will be displayed on the line.

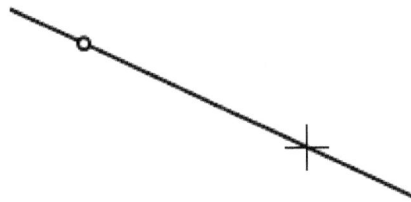

4. Move the cursor in the drawing window to position the line and click the cursor when you get the line in the desired orientation.

Lines are infinite and do not have points associated with them unless you specifically place one on the line.

Line is a modal command. You can continue making lines until you choose the select arrow or another drawing tool.

Active Axes -

The x and y axes have the properties of perpendicular infinite lines. When the crosshairs are positioned over an axis, the incidence symbol (bow tie) is

displayed at the intersection and the axis is highlighted.   When the

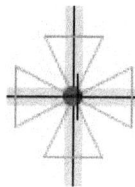

cursor is at the origin, both axes are highlighted . Points and End points of line segments can be placed directly on the axes without using the **Constrain / Incident** tool when the bow tie is displayed.

Drawing Vectors

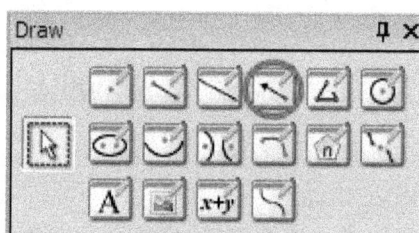

To add vectors to your geometry follow these steps:

1. Click the **Vector** icon in the **Draw** toolbox or select **Vector** from the **Draw** menu.

2. Position the cursor in the drawing window.

3. Click the mouse to place each endpoint.

Each vector is displayed with a letter label for each endpoint.

Drawing vectors is similar to drawing line segments, but vectors are constrained with coefficients of the form:

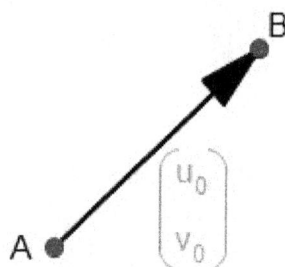

Vector is a modal command. You can continue making vectors until you choose the select arrow or another drawing tool.

Drawing Polygons

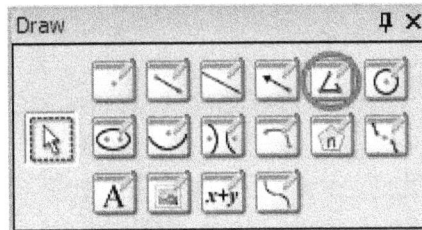

You can quickly create a multisided figure with these steps:

1. Click the **Polygon** icon ⬜ in the **Draw** toolbox or select **Polygon** from the **Draw** menu.

2. Position the cursor ┼ in the drawing window.

3. Move the cursor and click once to place each vertex.

   - As you create the sides of the polygon, each vertex is automatically assigned a letter name. You can change the label in Select ⬜ mode.

   - When you create the last side of the polygon by clicking on the first vertex, the polygon will be filled.

   - To change the appearance of the polygon (color or style), select ⬜ it, right click, and choose **All Properties** from the context menu.

   - **Polygon** is a modal command. You can continue making polygons until you choose the select arrow or another drawing tool.

Drawing Circles

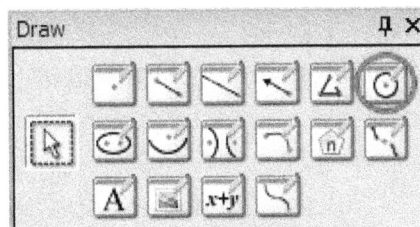

To add a circle to your diagram, follow these steps:

1. Click the **Circle** icon ⬜ in the **Draw** toolbox or select **Circle** from the **Draw** menu.

2. Move the cursor ✛ in the drawing window to the position of the center of the circle and click once.

3. Move the cursor to draw the circle in the desired size and click again.

Notice the circle is displayed with 2 points, the center and a point on the perimeter.

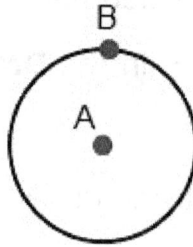

Circle is a modal command. You can continue making circles until you choose the select arrow or another drawing tool.

You can adjust the circle in Select ▨ mode.

Drawing Ellipses

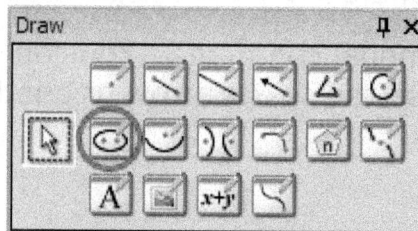

To create an ellipse in your diagram, follow these steps:

1. Click the **Ellipse** icon 🔘 in the **Draw** toolbox or select **Ellipse** from the **Draw** menu.

2. Move the cursor ✛ in the drawing window to the position of one focal point. Click to place the first focus point. Move the cursor and click again to place the second focal point.

3. Then move the cursor to open the ellipse to the desired shape and click the mouse a third time.

The ellipse will appear with three labeled points, the two foci and a point on the ellipse.

The **Ellipse** tool is a modal command. You can continue making ellipses until you choose the select arrow or another drawing tool.

Drawing Parabolas

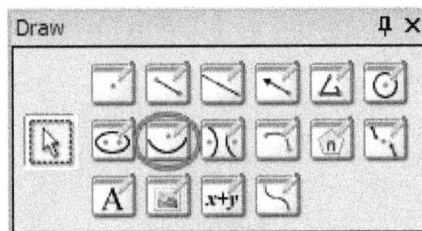

To create a parabola in your diagram, follow these steps:

1. Click the **Parabola** icon in the **Draw** toolbox or select **Parabola** from the **Draw** menu.

2. Move the cursor in the drawing window to the position of the parabola's vertex. Click and drag the mouse along the major axis. Release the mouse at the focus.

After sketching the general parabola, you can constrain it in the following ways:

1. Click the parabola and select **Implicit equation** from the **Constrain** toolbox and type or paste the formula.

$$X^2 \cdot a_0 + Y^2 \cdot |X^2 \cdot a + Y^2 \cdot b + 2 \cdot X \cdot Y \cdot c = 0| \quad e_0 + f_0 = 0$$

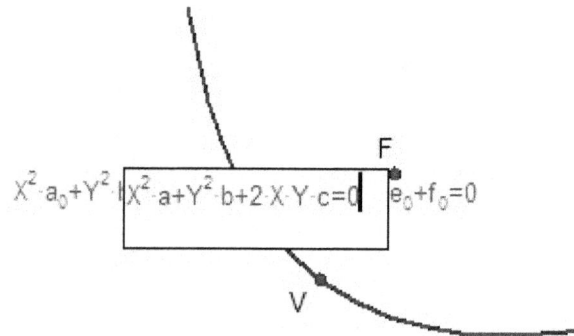

2. Constrain the vertex and focus points to some coordinate values.

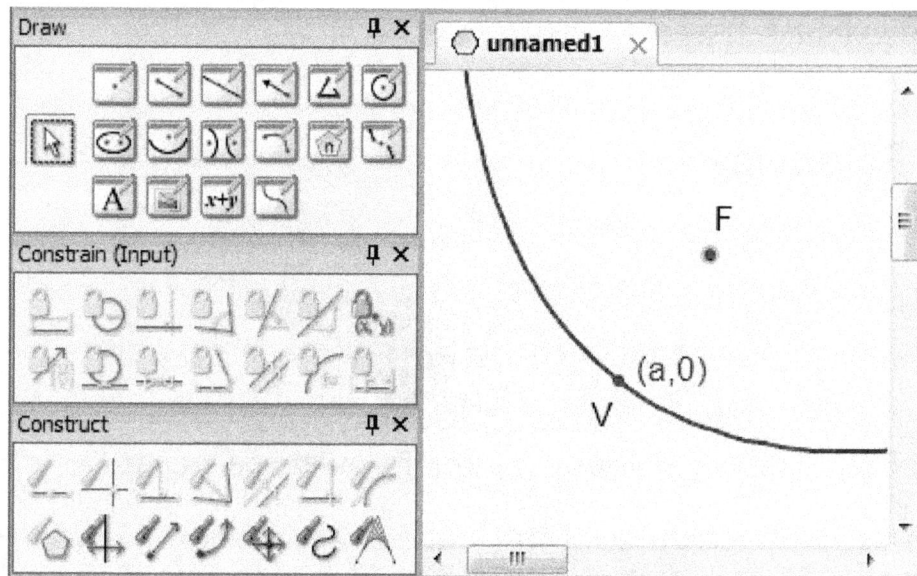

3. You can also adjust the parabola with the Translation, Rotation and Dilation tools.

Drawing Hyperbolas

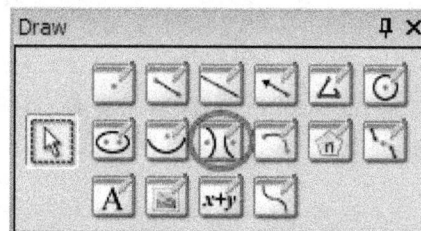

To create a hyperbola in your diagram, follow these steps:

1. Click the **Hyperbola** icon in the **Draw** toolbox or select **Hyperbola**

from the **Draw** menu.

2. Move the cursor ┼ in the drawing window to the position of one focal point. Click to place the first focus. Move the cursor and click again to place the second focal point.

3. Then move the cursor to open the hyperbola to the desired shape and click the mouse a third time.

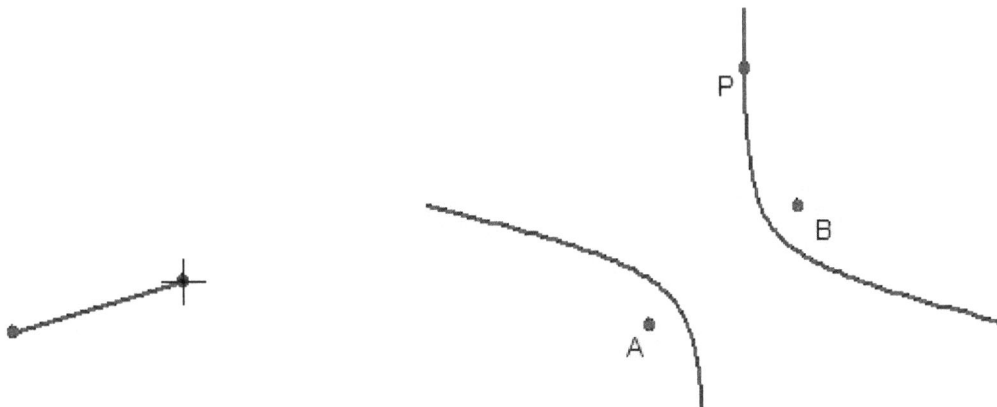

The hyperbola will appear with three labeled points, the two foci and a point on the hyperbola.

The **Hyperbola** tool is a modal command. You can continue making hyperbolas until you choose the select arrow or another drawing tool.

Drawing Arcs

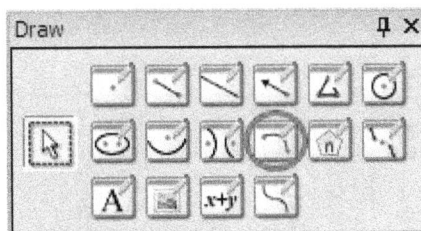

Arcs can be placed on any of the conics - circle, ellipse, parabola, hyperbola - or any function. Points are automatically placed at the ends of the arc. Here are the steps:

1. First draw the conic or function which will be the basis for the arc.

2. Choose the Arc tool and move the cursor over the section of the existing curve where the arc will be defined. Click and drag the cursor over the curve.

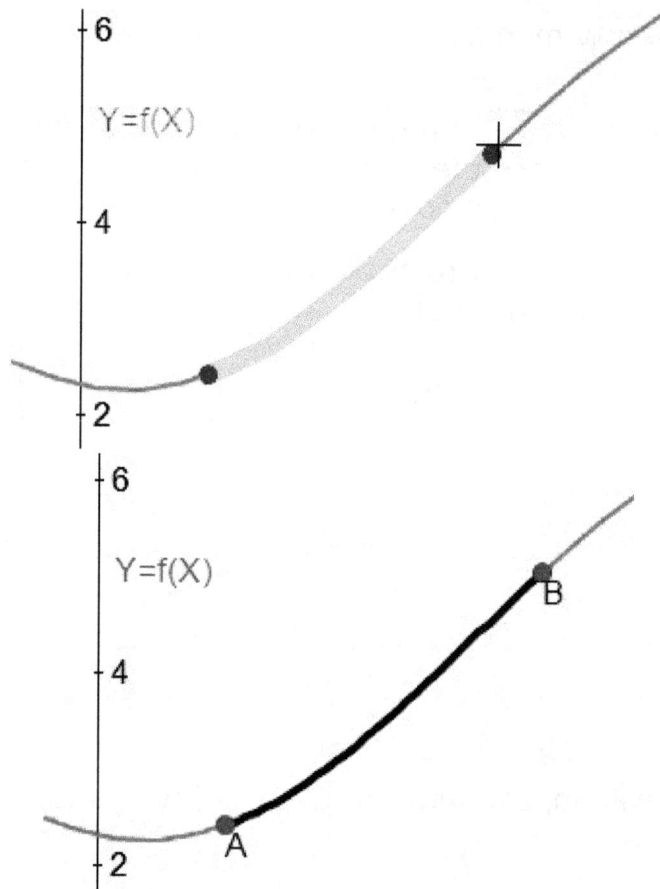

Endpoints are automatically inserted on the arc.

Drawing Regular Polygons

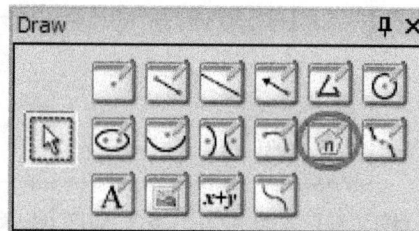

The **N-gon** tool lets you quickly draw any regular polygon. You can even work on problems where the number of sides is *n,* or whatever variable you choose.

Here are the steps:

1. Click the N-gon button in the Draw toolbox.

2. Similar to drawing a circle, position the cursor to place the center of the n-gon and click to the desired size. The n-gon at first appears to be a pentagon.

3. In the data entry box, enter the number of sides you want or a variable to represent the number of sides and press Enter.

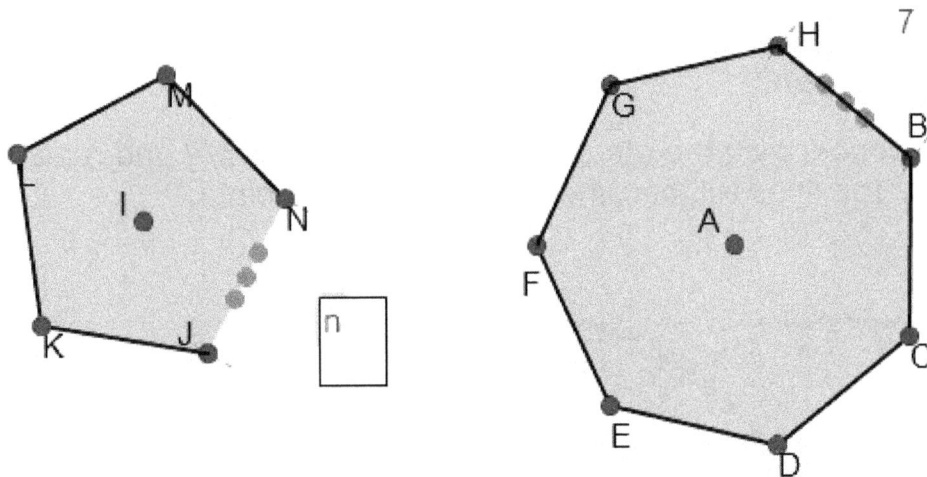

Curvilinear Polygons

The **Arc** drawing tool lets you make curvilinear polygons for which you can assign fill properties and find areas. There are some limitations, however. Since you can't construct point on two intersecting curves (except for circles), you have to connect curves with line segments. If you want to connect two arcs, you have to first connect them with a line and move it to the intersection like this:

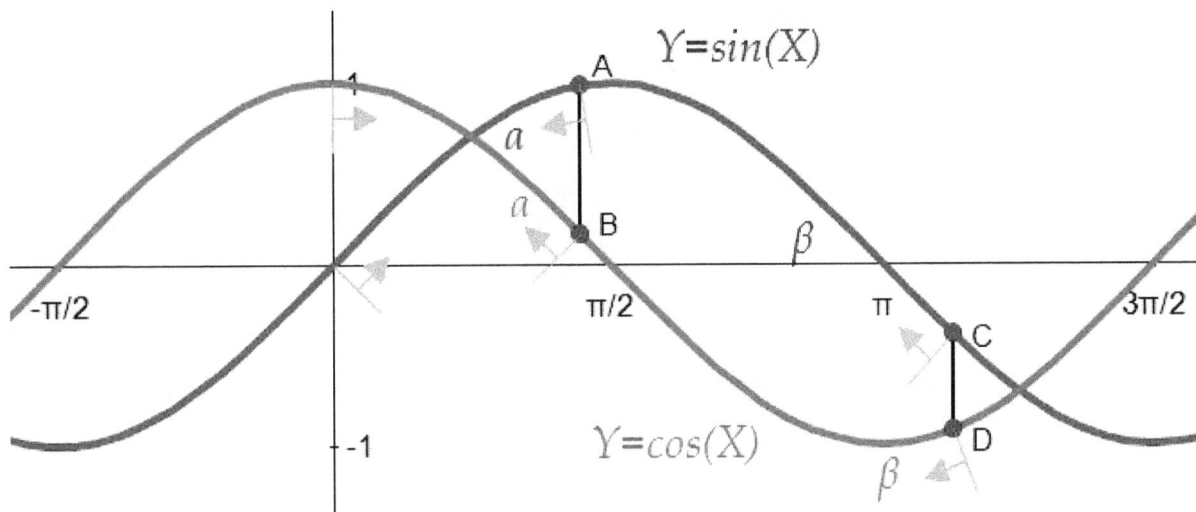

To make a curvilinear polygon of the intersecting functions here are the steps:

- Attach two lines.

- Make endpoints C and D β point proportional along the curves.

- Make endpoints A and B α point proportional along the curves.

- Draw the two arcs - select **Draw / Arc**, from C to A and D to B.

- Select the sides and arcs of the polygon in order and click **Construct / Polygon**.

- Double click the α's and replace them with π/4 and replace β with 5*π/4 . The lines will become the intersection points.

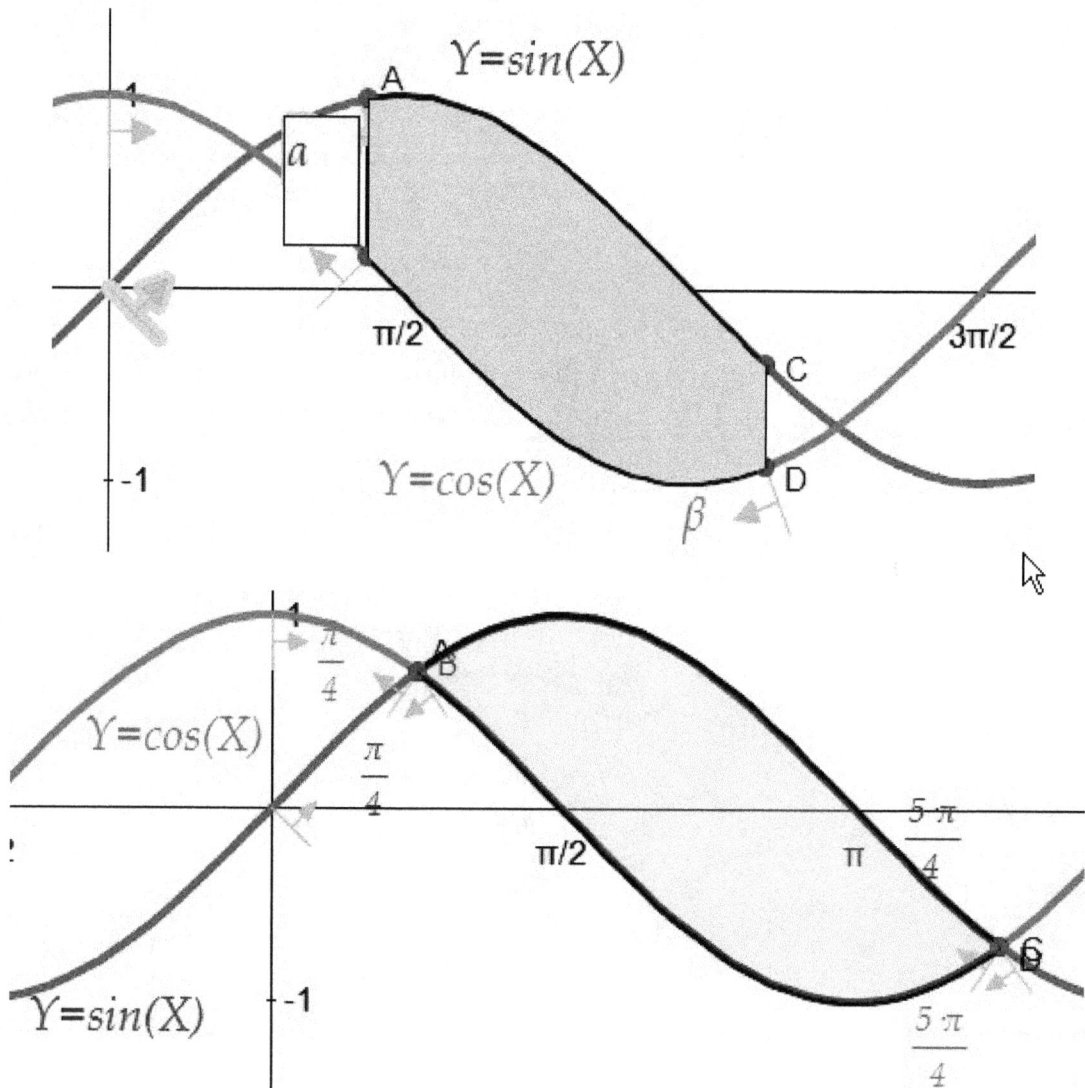

Drawing Curve Approximations

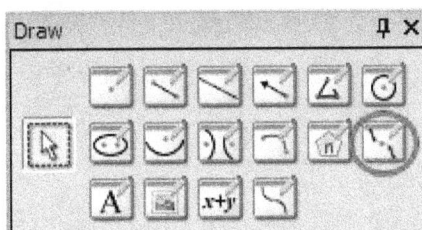

The **Curve Approximation** tool will insert a specified number of points and edges evenly spaced on a selected section of a curve or conic. This is a great tool for introducing problems using the Trapezoidal Method of integration. (Take a look at our Gx book, Calculus Explorations .)

Here are the steps:

1. Click **Draw / Curve Approximation** .

2. Select any function, circle, ellipse or parabola and drag the cursor over the curve.

3. In the data entry box type the number of points you want on the arc.

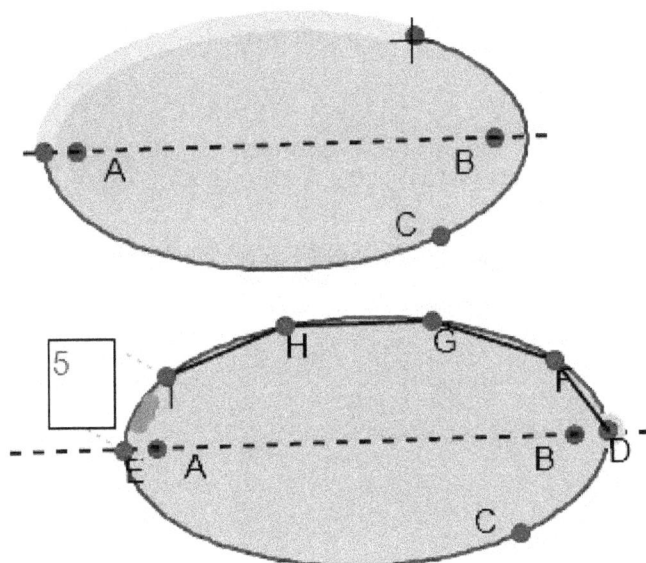

Note: It is best to draw your curve independent of existing points on the curve. Then connect other geometry to the approximation points. If you start or finish the arc with points lying on the curve (*e.g.* point C in the ellipse above) and later decide to delete the curve approximation, attached geometry may also get deleted.

Adding Text to the Drawing

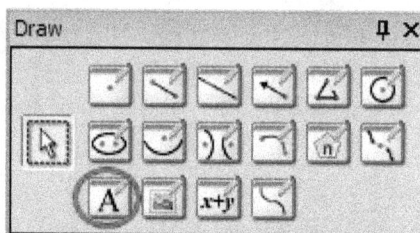

To add titles or other annotation to the drawing follow these steps:

1. From the **Draw** toolbox click the **Text** icon A or select **Text** from the **Draw** menu.

2. Position the text cursor ⁺A at the upper left corner where you would like your window of text located.

3. Click and drag to form your text box.

4. Enter and format your text in the Edit Text dialog.

Enter mathematical statements using the Annotation / Expression tool.

Inserting and Editing Text

In the **Edit Text** dialog you can enter and format the text that will be displayed in your defined text window.

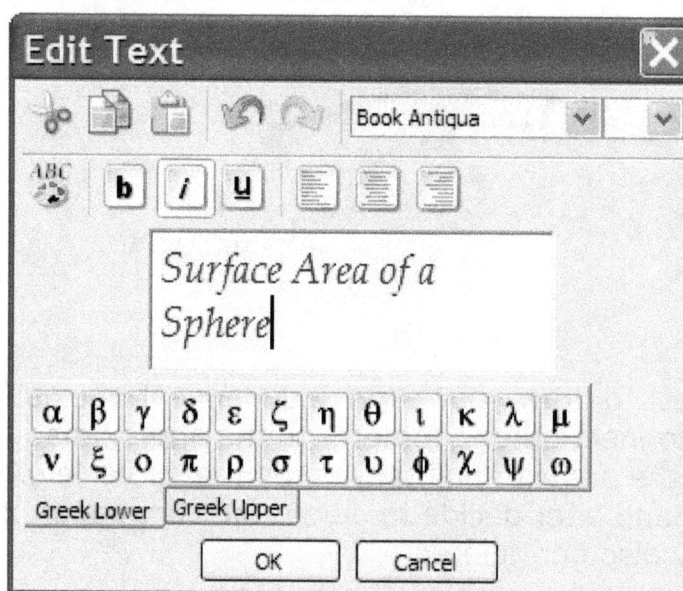

The default text formatting is set in **Edit/Preferences/Text & Pictures**.

Using Pictures in the Drawing

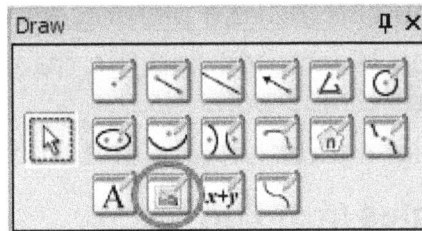

Liven up your examples with a picture or two, or use an image for reference points in your drawing. Here's how:

1. Click the **Picture** icon in the **Draw** toolbox or select **Picture** from the **Draw** menu.

2. Click and drag the cursor to delimit the area where you want to place the picture.

3. Find your image in the **Select Image File** dialog. Image formats include: **.bmp**, **.gif**, **.jpg**, **.pcx**, **.png**, and **.tif**

4. After entering your images, change to **Select** mode or choose another **Draw** command.

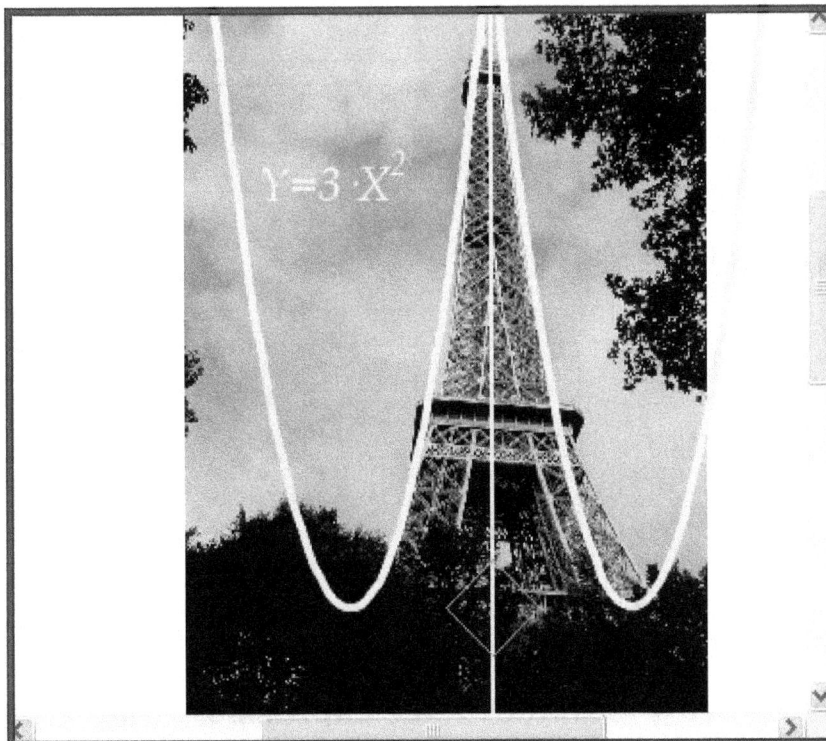

Pictures are always inserted under your drawing objects, so you can add a picture at any time.

In **Select** ⌖ mode you can move a picture, as with any drawing object. The inserted picture can be rotated and the transparency level can be set in the **Display Properties** dialog box.

- Right click on a highlighted picture and then select **All Properties.**

- Or select **Properties** from **Edit** menu.

Display Properties	
⊟ Picture	
Visibility Condition	
Pinned	☑
Angle In Degrees (-359 to 359)	0
Transparency (0% to 100%)	0

OK   Cancel

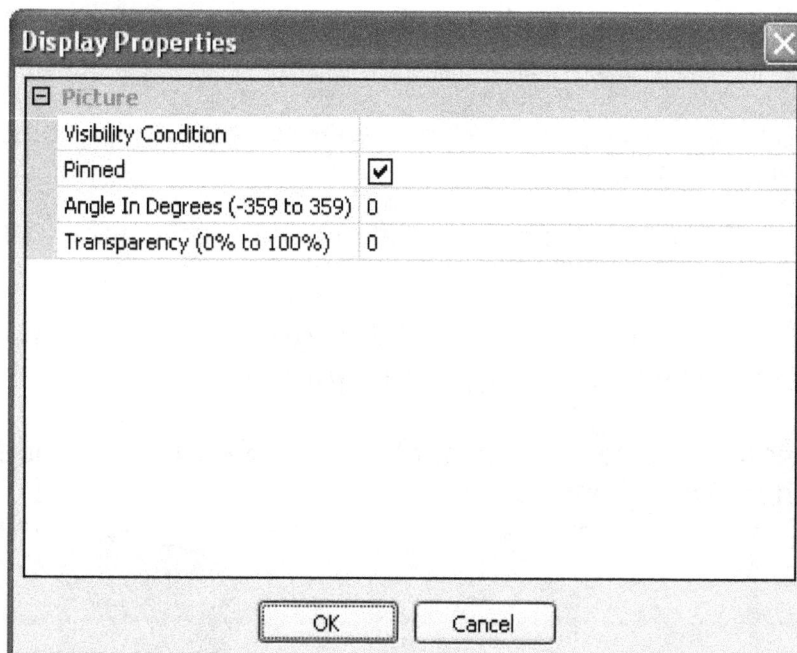

Adding Expressions

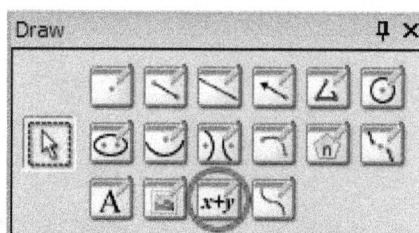

You can type an algebraic expression in the drawing window and Math Illustrations will solve it with whatever information it has available. Here are the steps:

1. From the **Draw** toolbox click the **Expression** icon or select **Expression** from the **Draw** menu.

2. Move the expression cursor to the position where you want it to

appear in the drawing window and click to display the data entry box.

3. Enter the expression using numbers, variables. Use the Symbols toolbox to help you enter mathematics.

Here's an example:

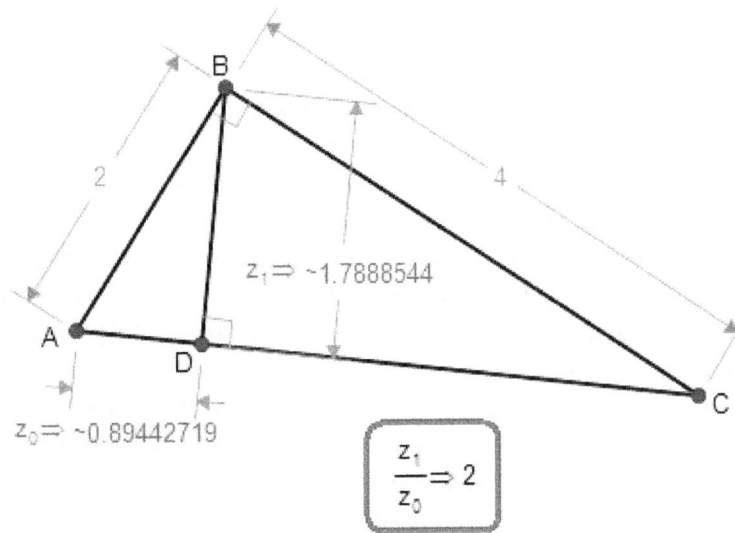

Creating Functions

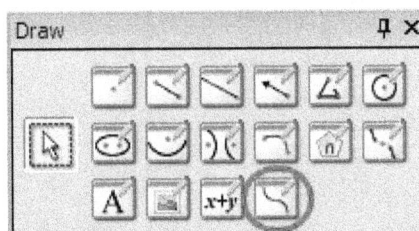

1. Click the **Function** icon in the **Draw** toolbox or select **Function** from the **Draw** menu.

2. Click the drop-down list button to select the function type that you want to use in your drawing.

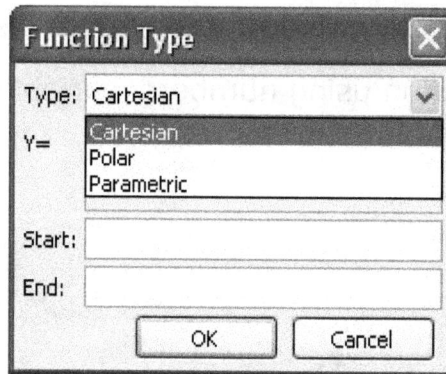

(!) Important Note for Trig Functions: Remember that in order for geometry to be displayed accurately on the screen, *Geometry Expressions* uses equal scaling in the *X* and *Y* directions. So in order for your trig functions to be displayed properly you must be in **Radians** mode - check the window on the bottom right side of the screen.

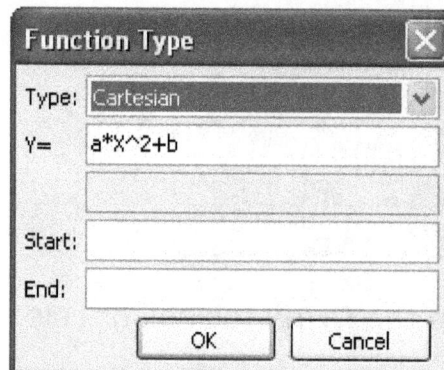

. No matter what the mode, you still have a choice in labeling the *X* axis from the **Context** menu, **Axis Properties / Units**.

Cartesian Function

When you select **Cartesian** from the **Function Type** dialog, the next line contains a general form of the function in terms of Y.

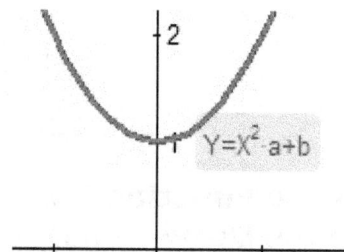

You can define a domain of the function by enter values for Start and End. If you want to draw function with indefinite domain, leave these fields blank.

You can define this function in these ways:

* Modify the formula directly:

Function Type

Type: Cartesian

Y= a*X^2+b*X^|

Start:

End:

OK   Cancel

$Y=X^3 \cdot a + c + b \cdot X^2$

-2   -1   1   2
2

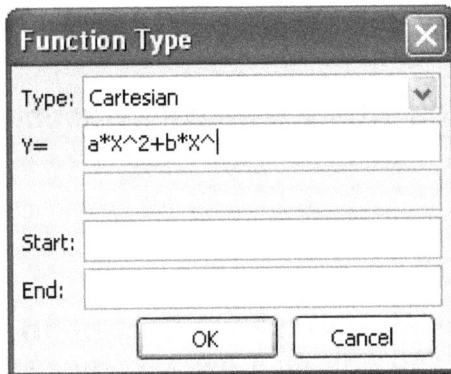

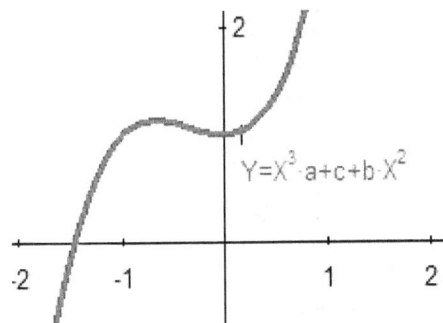

- Modify the function interactively using "handles". In the drawing window, click the function to select it. Click and drag it and a circle appears on the curve. This handle represents one of the variables in the equation. You can click and drag this handle around the drawing. Click and drag another place on the function and another moveable handle appears if there is another variable in the equation.

This feature is a wonderful way to understand exactly how the equation represents the function.

In the first example, a click of the curve gives you the *b* handle, the *y*-intercept, and lets you drag the function up and down. Click and drag another place on the curve and you get the *a* handle to change the shape of the curve.

$Y=X^2 \cdot a + b$

$Y=X^2 \cdot a + b$

Variables

Variables	Functions	
Name	Value	Locked
a	-0.20542591	-
b	3.5221638	-

Notice the **Variable** toolbox displays the changing values as you move the handles.

- Modify the function after it's drawn by double clicking the function tag and changing it in the edit box
- Modify the domain of the function after it's drawn by double clicking the curve.

The Generic Function f(X)

To use the generic form of a function, Y=f(X), select **Cartesian** from the **Function Type** dialog, enter f(X) in the edit window and click <u>OK</u>:

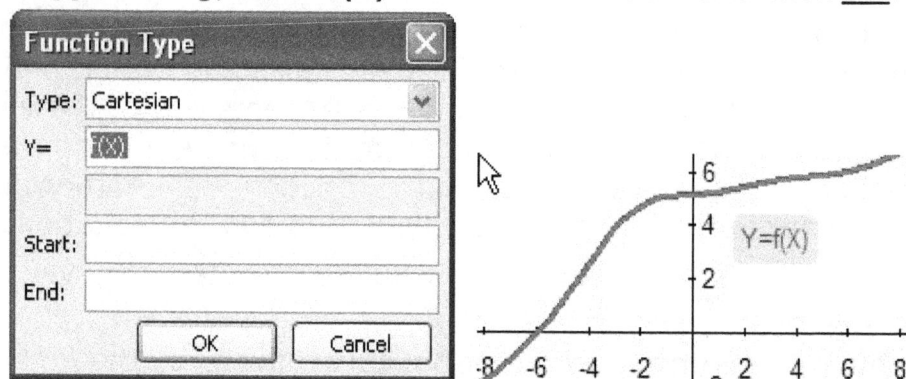

Click the **Functions** tab of the **Variables** toolbox to see the values used in the function.

You can define this function in these ways:

- Modify the edit line at the bottom for the **Functions** tab. (Use your keyboard arrow keys to move through the whole function.)

- Modify the function interactively using "handles". In the drawing window, click the function to select it. Click and drag it and a circle appears on the curve. This handle represents one of the variables in the equation. You can click and drag a handle around the drawing to change the curve. Click and drag another place on the curve and another moveable handle appears if there is another function variable in the equation. In the general function in this example we have 5 possible handles: f(a), f(b), f(k), f(u), and f(v).

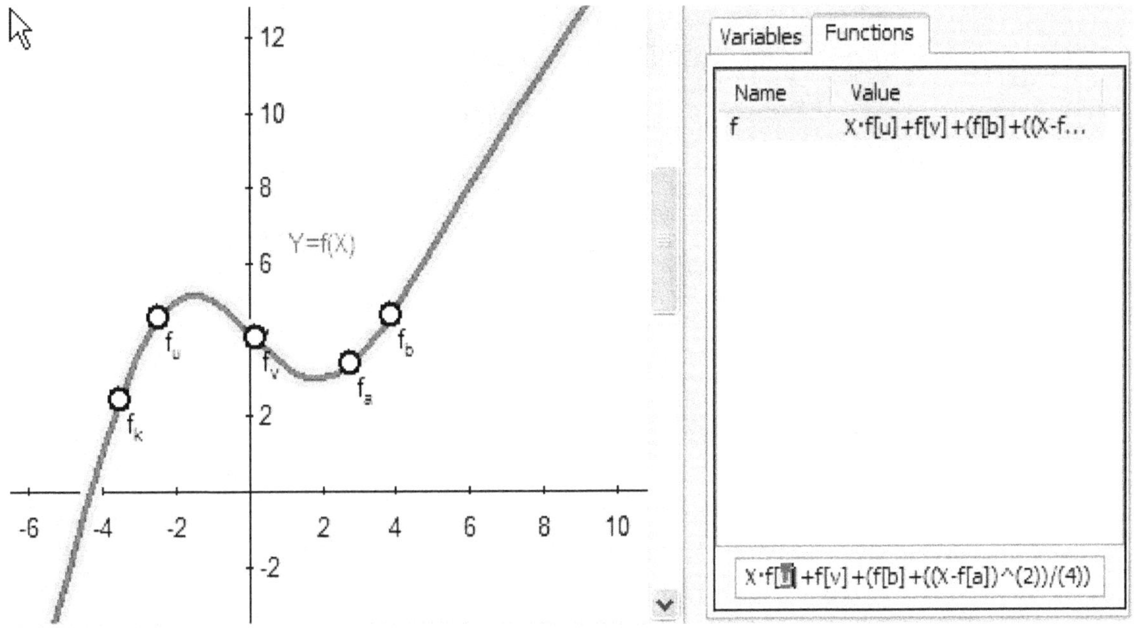
Y=f(X)
Variables
Functions
Name
Value
f
X·f[u]+f[v]+(f[b]+((X-f...

Polar Function

When you select **Polar** from the **Function Type** dialog, the next line contains the general form of the function in terms of the radial coordinate, r and the polar angle, T.

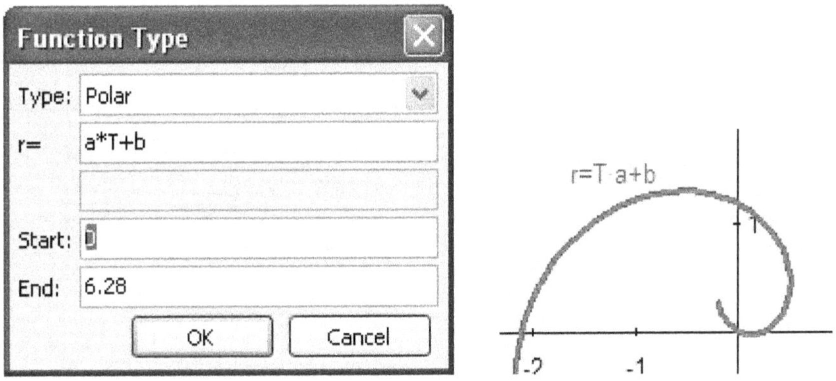
Function Type
Type: Polar
r= a*T+b
Start:
End: 6.28
OK
Cancel
r=T·a+b

You can define this function in these ways:

- Modify the formula and the curve domain directly:

Function Type

Type: Polar

r=   a*T^3+b

Start: 0

End: 6.28

OK   Cancel

$r=T^3 \cdot a+b$

10

0   -20   -10

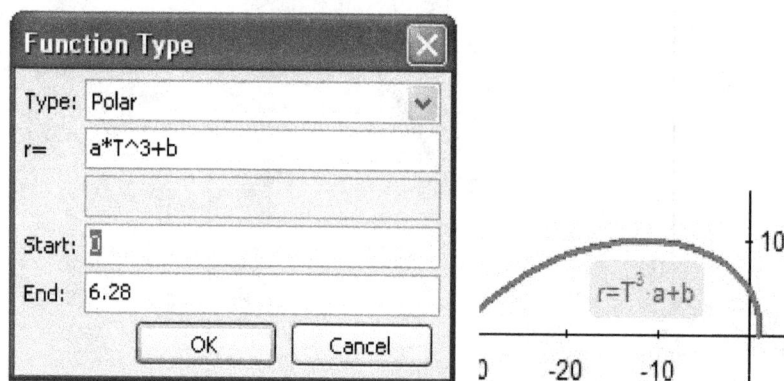

- Modify the function interactively using "handles". In the drawing window, click the function to select it. Click and drag it and a circle appears on the curve. This handle represents one of the variables in the equation. You can click and drag this handle around the drawing. Click and drag another place on the function and another moveable handle appears if there is another variable in the equation.

  This feature is a wonderful way to understand exactly how the equation represents the function.

  In the example above, the second click gives you the *a* handle; click and drag another place on the curve and you get the *b* handle to change the shape of the curve.

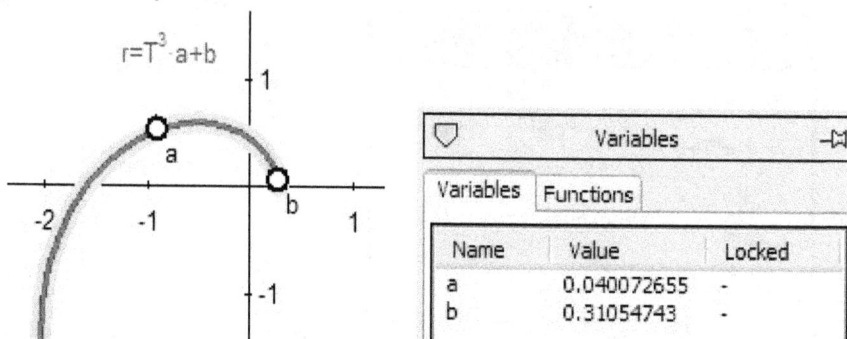

$r=T^3 \cdot a+b$

1

a

-2   -1   b   1

-1

Variables

Variables | Functions

Name	Value	Locked
a	0.040072655	-
b	0.31054743	-

Notice the **Variable** toolbox displays the changing values as you move the handles.

- Modify the function after it's drawn by double clicking the function tag.

- Modify the domain of the function after it's drawn by double clicking the curve:

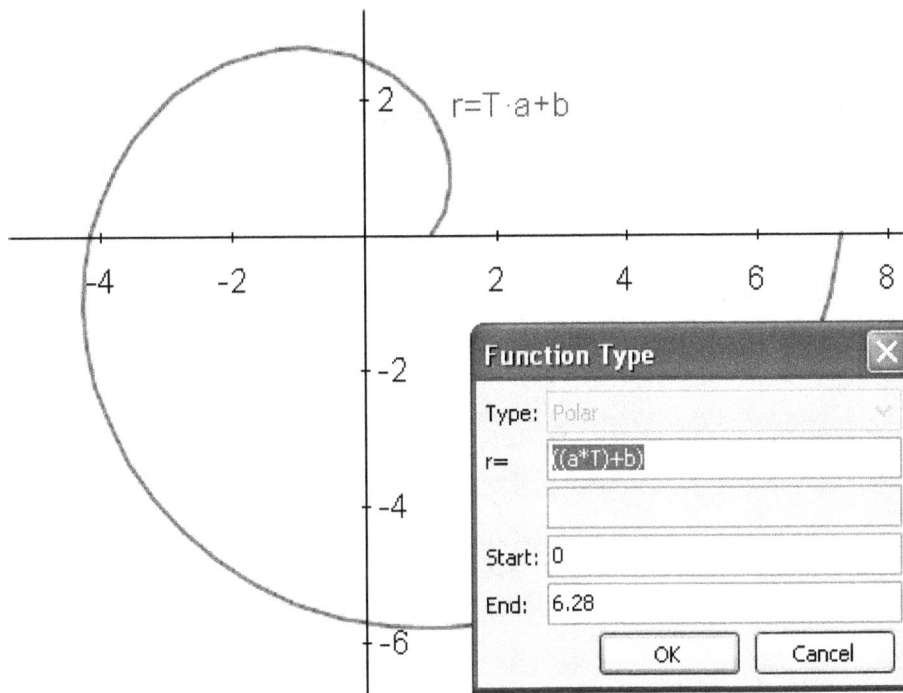

r=T·a+b

Function Type

Type:	Polar
r=	((a*T)+b)
Start:	0
End:	6.28

OK   Cancel

Piecewise Function

A piecewise function or expression can be created using the **Piecewise** symbol:

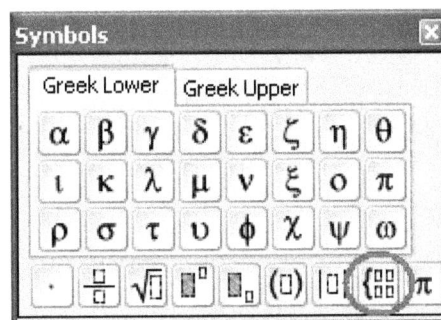

or the built-in function - piecewise({expression1, domain1},{expression2, domain2}...,{last expression, otherwise}). The reserved word, "otherwise" is an option available for the last condition.

Here we show how to enter the sequence of values, followed by conditions:

- Create a function ▱.
- Double click the equation to edit.

- From the edit box click the **Piecewise** icon in the **Symbols** toolbox. Four small gray boxes will appear, including the one containing the original equation.

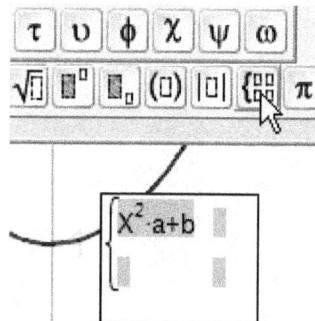

- Fill in your piecewise parameters - equations in the left column and their domain in the right column. To expand the function, click the **Piecewise** icon again for an additional row.

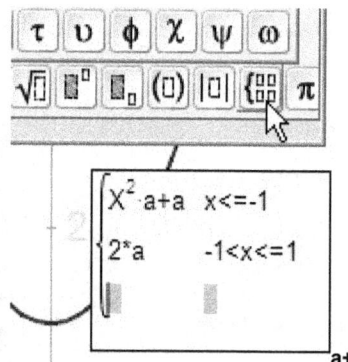

Here's the finished piecewise function:

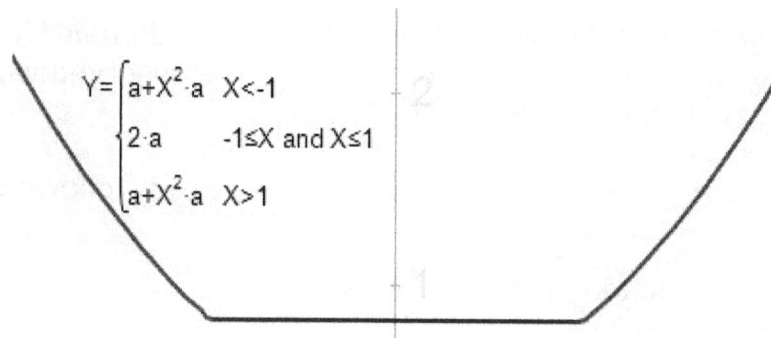

$$Y = \begin{cases} a + X^2 \cdot a & X < -1 \\ 2 \cdot a & -1 \le X \text{ and } X \le 1 \\ a + X^2 \cdot a & X > 1 \end{cases}$$

Parametric Function

When you select **Parametric** from the **Function Type** dialog, the next line contains the general form of the function in terms of X and Y and a parameter, T.

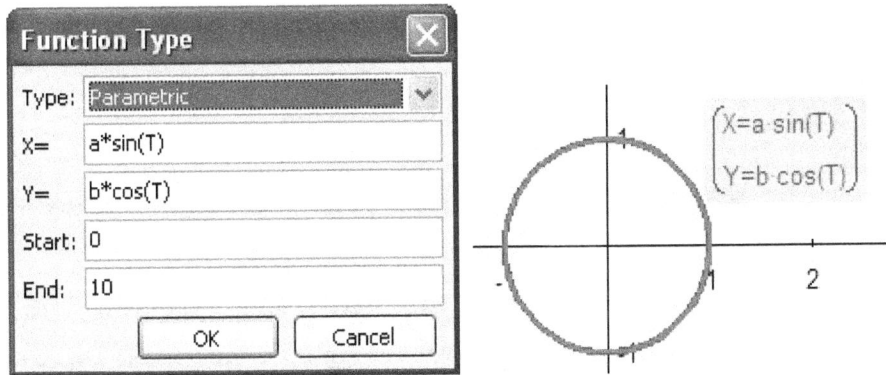

You can define this function in these ways:

- Modify the formula and its domain directly:

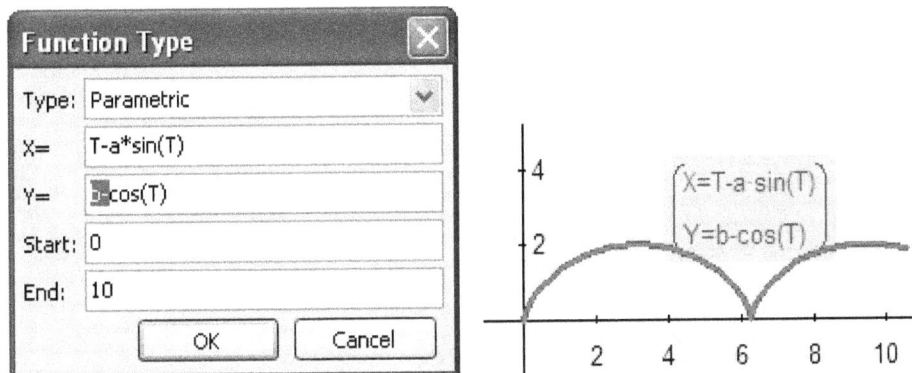

- Modify the function interactively using "handles". In the drawing window, click the function to select it. Click and drag it and a circle appears on the curve. This handle represents one of the variables in the equation. You can click and drag this handle around the drawing. Click and drag another place on the function and another moveable handle appears if there is another variable in the equation.

This feature is a wonderful way to understand exactly how the equation represents the function.

In the example above, X=T-$a$sin(T), Y=b-cos(T), a click and drag gives you the b handle, and lets you drag the function up and down. Click and drag another place on the curve and you get the a handle to change the shape of the curve.

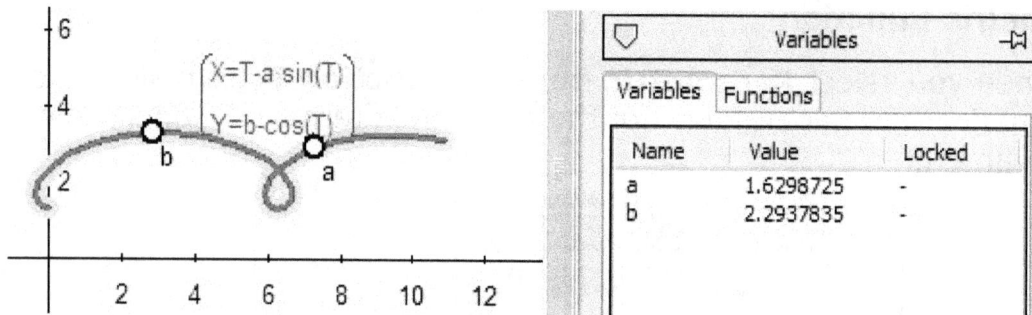

Notice the **Variable** toolbox displays the changing values as you move the handles.

- Modify the function after it's drawn by double clicking the function tag.

- Modify the domain of the function after it's drawn by double clicking the curve:

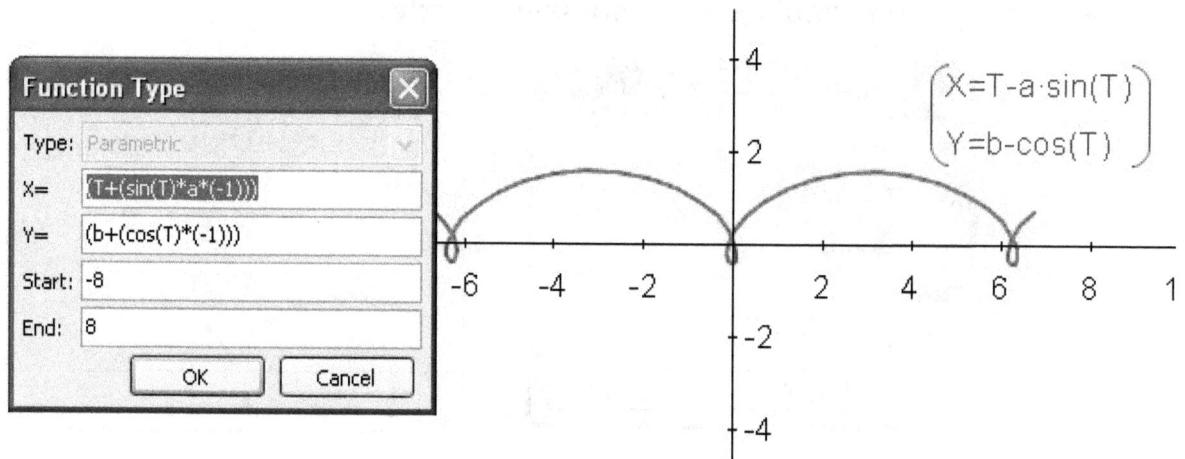

Piecewise Parametric Example

Any of the functions available in *Geometry Expressions* can be piecewise, including parametric functions. Take this square for example.

1. Click the **Draw / Function** tool
2. Select **Parametric** from the drop down <u>Type</u> window.

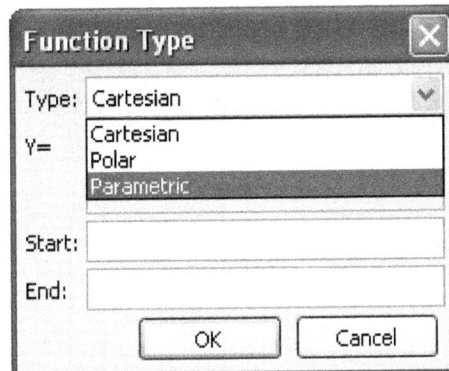

3. Enter the first value (side) for the square and the range for the parameter values -

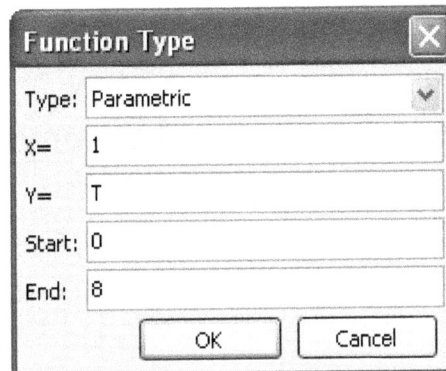

Hit enter. Don't worry about the shape of the function, it's not defined yet.

4. Double-click the function and select the x value

5. Click the **Piecewise** icon

6. Enter the x values for each side of the square.   When you run out of gray boxes, click the **Piecewise** icon again to get another row. -

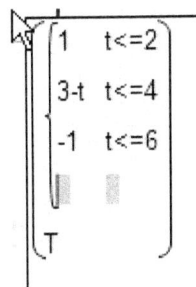

7. Now do the same for the y value, select the T parameter and click the **Piecewise** icon 3 times -

$$\begin{cases} 1 & t<=2 \\ 3\text{-}t & t<=4 \\ \text{-}1 & t<=6 \\ t\text{-}7 & t<=8 \\ T & \\ & \\ & \\ & \end{cases}$$

Use the arrow keys on your keyboard or your mouse to move to the next gray box.

Here is the function:

$$X = \begin{cases} 1 & T\leq2 \\ 3\text{-}T & T\leq4 \\ \text{-}1 & T\leq6 \\ \text{-}7\text{+}T & T\leq8 \end{cases}$$
$$Y = \begin{cases} T & T\leq2 \\ 2 & T\leq4 \\ 6\text{-}T & T\leq6 \\ 0 & T\leq8 \end{cases}$$

The Selection Arrow

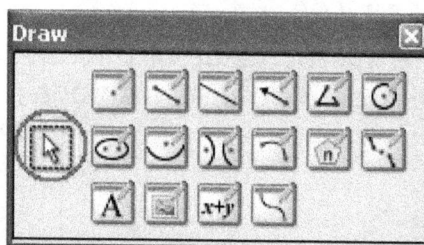

When you are finished with the drawing functions, click the selection arrow to enable other functions or adjust your drawing.

With the arrow, you must first select elements of the drawing in order to enter constraints and constructions and to output calculations.

Many of the tools require you to select multiple objects.

The **Selection** arrow is also available form the icon bar at the top of the screen.

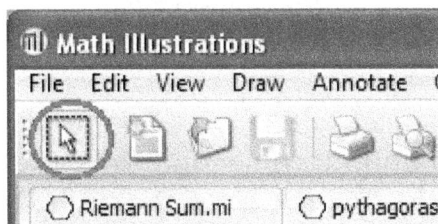

Selecting Multiple Objects

Many of the **Constrain**, **Construct**, and **Calculate** tools require that you select more than one object.

To select more than one object:

- Click the selection arrow on the tool bar.

- Hold down the CTRL or SHIFT key as you click the objects you want to select.

If you click the wrong object while holding down the CTRL key, click it again and it will become unselected.

If all of the objects you wish to select fall within a rectangular region, you can use the selection box:

- Click on the selection arrow on the tool bar or the **Drawing** toolbox.

- Draw a rectangle around all of the objects by clicking and dragging.

If you are using *Math Illustrations* on a *Smartboard*, turn on Smartboard mode selection. In this mode selection accumulates. Clicking off of an item does not clear the selection. Clicking on an item a second time unselects that item.

Smartboard Mode

If you are using *Math Illustrations* on a *Smartboard*, it's a good idea to switch to Smartboard Mode.   Under **Preferences / Selection**, choose Smartboard mode.

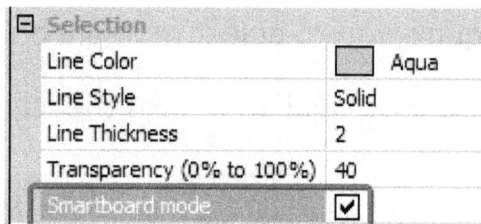

When check, selection accumulates. Clicking off an item does not clear the selection. Clicking on an item a second time unselects that item. To clear an entire selection, select **Clear Selection** under the **Edit** menu.

Adjusting the Drawing

Click the Select arrow , either from the **Drawing** toolbox or the icon bar, to move, rotate or delete selected object(s) in the drawing.

You can change a constraint value, annotation or label by double clicking it, retyping the value or variable and then pressing the enter key.

The **Scale, Zoom** and **Move / Pan** functions have a mouse shortcut:

- The scroll wheel on your mouse can be used to **Scale** the drawing up or down. Hold down the ctrl key while moving the wheel and the operation becomes a **Zoom.**

- Right-click and drag the mouse anywhere in the drawing window to **Move** or **Pan.**

The Move geometry icon is a modal command. It stays active until you select or choose another mode (e.g. any Draw tool).

Constraints

Using Drawing Constraints

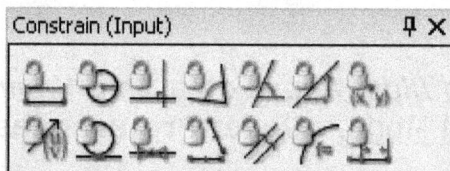

After sketching the geometry of a problem, constrain it with measurements, coordinates and implicit equations in real or symbolic terms. The drawing responds automatically to the assigned input constraints. *Math Illustrations* will automatically add any constraints you leave out.

Since annotations may look identical to constraints, use the icon, , to **Distinguish Constraints / Annotations**. The icon is a toggle; to turn off the marks, click it again.

Initially, all the constraints in the toolbox are inactive. You must first select the parts for your drawing that will be constrained. Constraint choices are listed below along with the drawing elements that must be preselected. Be careful when selecting geometry objects, if extra things are selected that are not related to the constraint (like other constraints) the constraints will remain inactive. This can happen by mistake, especially when using the selection box tool.

	Constraint	**Preselected Objects**
	Distance / Length	Two of any combination of points, lines, line segments, vectors, or polygon sides.
	Radius	A circle
	Perpendicular	Two of any lines, segments, vectors, or polygon sides.
	Angle	Two of any lines, segments, vectors, or polygon sides.
	Direction	A line, segment, vector, or polygon side.
	Slope	A line, segment, vector, or polygon side.
	Coordinate	A point
	Coefficients	A vector
	Tangent	A circle or locus and a line, segment, vector, or polygon side.

	Incident	A point and a line, segment, vector, polygon side, circle or locus.
	Congruent	Two or three of any line segment, vector, or polygon side.
	Parallel	Two or three of any line segment, vector, or polygon side.
	Implicit Equation	A circle, line, segment, vector, or polygon side.
	Point Proportional Along Curve	A point and a line, segment, vector, polygon side, or locus

Occasionally you may try to add too many constraints to the geometry, causing a conflict. The system will help you correct this problem in the Resolve Constraint Conflict dialog.

Changing a Constraint

To change a constraint, double click it with the selection arrow, [ ] retype the value or variable and press the enter key.

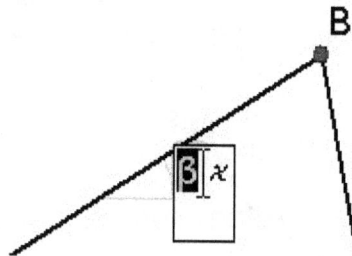

Constraint Conflicts

If you enter a constraint for some geometry which is already constrained by another constrained object you will see a message like the one below.

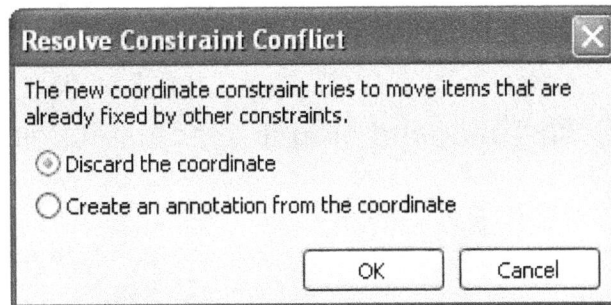

In this case, the **Coordinate** constraint was already determined by the other two sides and an angle constraint on the triangle. There are two ways of dealing with this problem:

1. Hit the **Cancel** button (or select the first option, "Discard the coordinate", and then click OK) to leave the drawing as it was without the new constraint.

2. Select the second option, "Create an annotation from the coordinate", and then click OK, the coordinate constraint become annotation coordinates. It just show on the diagram and has no effect on the geometry engine.

Distance / Length Constraint

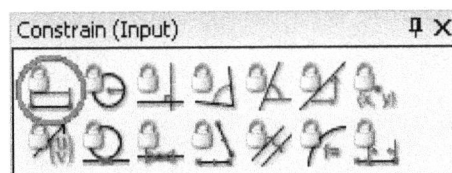

The **Distance / Length** constraint lets you specify the following dimensions:

- Length of a line segment, vector, or polygon side

- Distance between two points or a point and any one of the line types listed above.

- To enter a constraint:

    1. Select the appropriate drawing object(s). When you make your selection, the **Distance / Length** icon will light up .

    2. Click the icon, enter the constraint value, either real or symbolic, and

press enter. You can press enter without typing a value to accept the system's default value.

You can click the constraint and drag it to adjust its placement on the drawing.

Radius Constraint

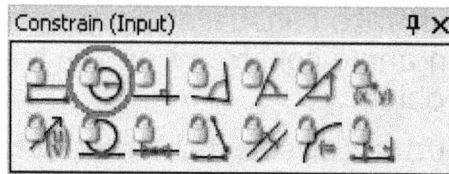

To specify the radius of a circle:

1. From select mode [cursor icon], click the circle. The circle will be highlighted as well as the icon [icon].

2. Click the **Radius** icon, enter the constraint value, either real or symbolic, and press enter. You can press enter without typing a value to let the system insert a variable name.

You can click the constraint and drag it to adjust its placement on the drawing.

Perpendicular Constraint

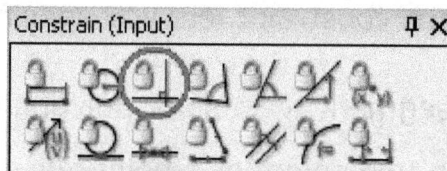

Any two of lines, segments, vectors or polygon sides can be constrained to be perpendicular with these steps:

1. Select [cursor icon] two from the line types listed above.

2. Click the **Perpendicular** icon [icon] .

The lines are redrawn and the perpendicular constraint is attached.

Angle Constraint

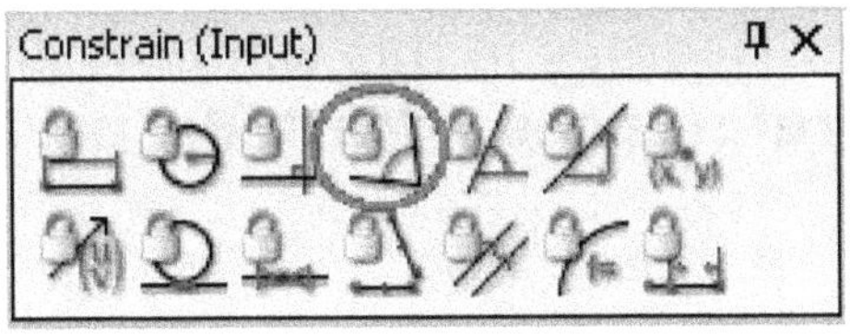

Any two of lines, segments, vectors or polygon sides can be constrained with an angle value or variable name with these steps:

1. Select 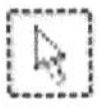 two from the line types listed above.
2. Click the **Angle** icon 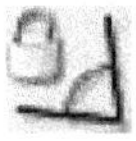 .
3. Enter the constraint, real or symbolic. If you enter a real value, the lines will be adjusted to reflect the constraint.

Which Side to Constrain?

Sometimes when identifying angles, the constraint falls on the wrong one. In the example below, we wanted BDC, not BDA. Just click the cursor over the constraint arrow and drag it to the other side, then release the mouse button - done!

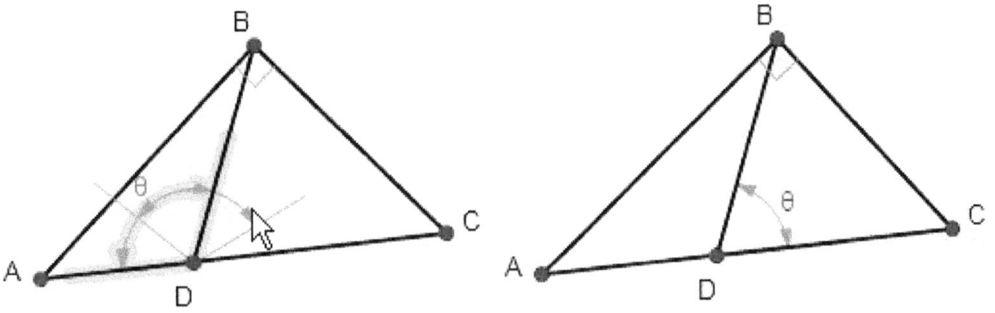

To get the reflex angle, hold the *Ctrl* key while dragging the angle symbol.

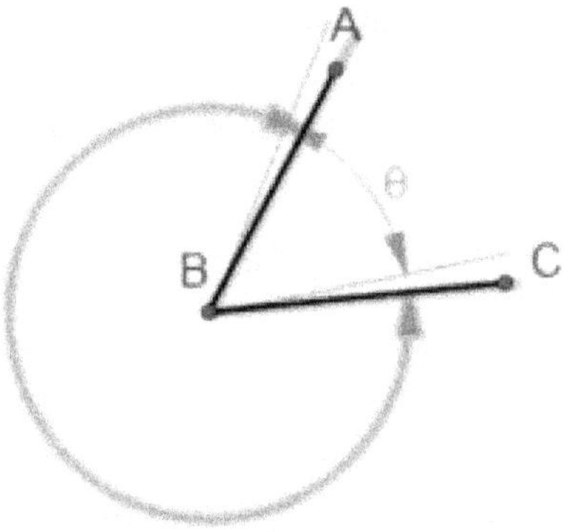

Note: The angular units are displayed in the lower right of the screen. You can choose Radians or Degrees in this window, or set the default Angle Mode in the **Edit / Preferences / Math Properties** menu, **Math** settings group.

Direction Constraint

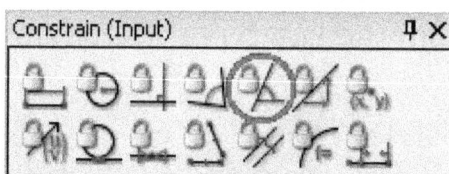

Constrain any of the line types; line, line segment, vector, or polygon side, to a direction measured from the horizontal.

1. Select [⬚] one of the line types listed above.

2. Click the **Direction** icon .

3. Enter the constraint, real or symbolic. If you enter a real value, the line will be adjusted to reflect the constraint.

Note: The angular units are displayed in the lower right of the screen. Change the default (Degrees or Radians) in the **Edit / Preferences** menu.

Slope Constraint

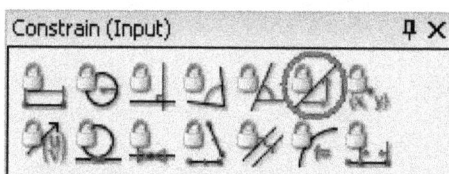

Specify a slope for any of the line types; line, line segment, vector, or polygon side.

1. Select [⬚] one of the line types listed above.

2. Click the **Slope** icon .

3. Enter the constraint, real or symbolic. If you enter a real value, the line will be adjusted to reflect the constraint.

Coordinate Constraint

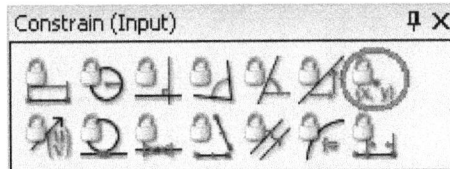

Constrain (Input)

You can give coordinates to any point in your drawing:

1. Select a point.

2. Click the **Coordinate** icon .

3. Enter the constraint, real or symbolic. If you enter a real value, the line will be adjusted to reflect the constraint, even if the coordinate axes are not displayed.

To change the coordinates shown, double click and type over the highlighted value in the data entry box.

Constraining Vector Coeffecients

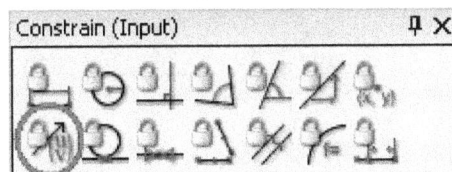

Constrain (Input)

You can specify coefficients for a vector with the following steps:

1. Select a vector.

2. Click the coefficients icon .

3. Enter the coefficients separated by a comma.

Note: Don't forget the parentheses or an error message appears.

Tangent Constraint

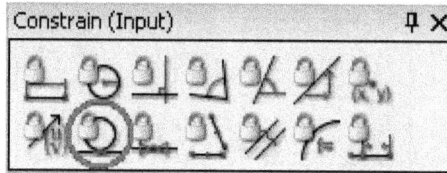

Any of the line types; line, line segment, vector, or polygon side can be made tangent to a circle or locus with these steps:

1. Select ⬚ a line of the types listed above and the circle or locus.

2. Click the **Tangent** icon ⬚ from the **Constrain** tool box or select **Tangent** from the **Constrain** menu.

The line and curve immediately become tangent.

Incident Constraint

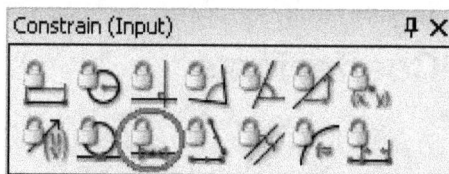

Constrain a point to be incident to any other geometry; line, segment, vector, polygon side, circle or locus with these steps:

1. Select ⬚ the point and the other geometry listed above.

2. Click the **Incident** icon ⬚ from the **Constrain** toolbox, or select **Incident** from the **Constrain** menu.

The point is moved to meet the line or curve, or the extension of the line.

Below is an example of the latter, point D is moved to lie on the extension of line segment AB.

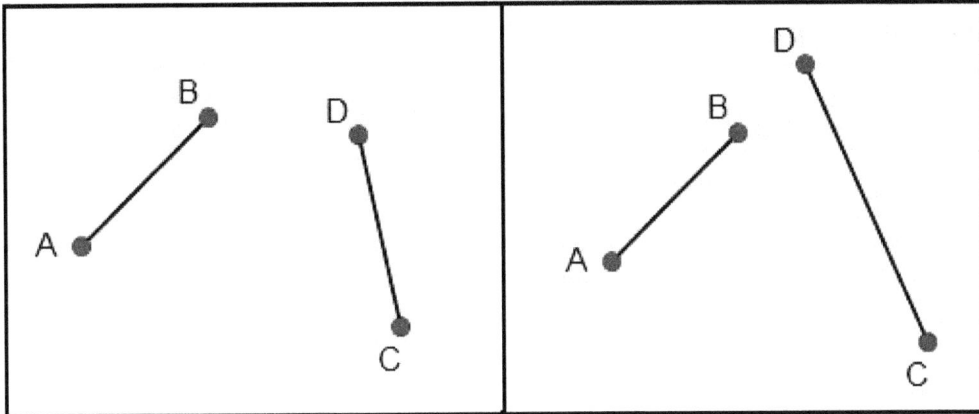

If you select the point or the line, incidence is indicated by a bowtie around the point:

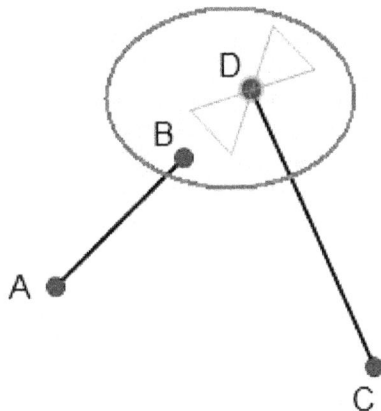

Congruent Constraint

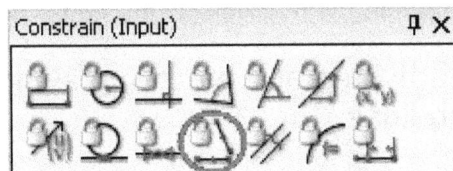

Constrain two or three of any of these geometry types: line segments, vectors, or polygon sides, to be congruent with these steps:

1. Select [cursor icon] two line segments.

2. Click the **Congruent** icon [icon] from the **Constrain** toolbox, or select **Congruent** from the **Constrain** menu.

You will see matching congruency lines on the selected segments and a length will be adjusted.

Parallel Constraint

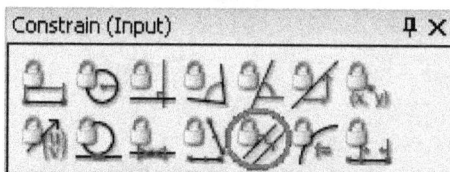

Any two or three of the linear geometry types can be made parallel line, segment, vector, or polygon side.

1. Select [⌖] two from the types listed above.

2. Click the **Parallel** icon ⫽ from the **Constrain** toolbox, or select **Parallel** from the **Constrain** menu.

The geometry will be adjusted and matching symbols appear on the selected lines.

Implicit Equation Constraint

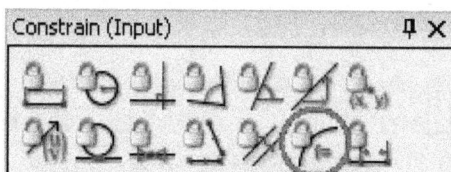

You can use symbolic variables to constrain geometry with an implicit equation. Lines, line segments, polygon sides, vectors and circles and conics can all be constrained with implicit equations.

1. Select ⌶ the geometry.

2. Click the **Implicit Equation** icon 🔒 from the **Constrain** toolbox, or select **Implicit Equation** from the **Constrain** menu.

An input window will open next to the geometry you selected. Highlighted in the window is a generic equation for the selected object; for a line, an equation like - $XA_0+YB_0+C_0$ Y== 0 might appear. You can edit the equation with different variable names or coefficients as you like or enter a totally different equation. For example, we've given the line the equation:

$Y=X$.

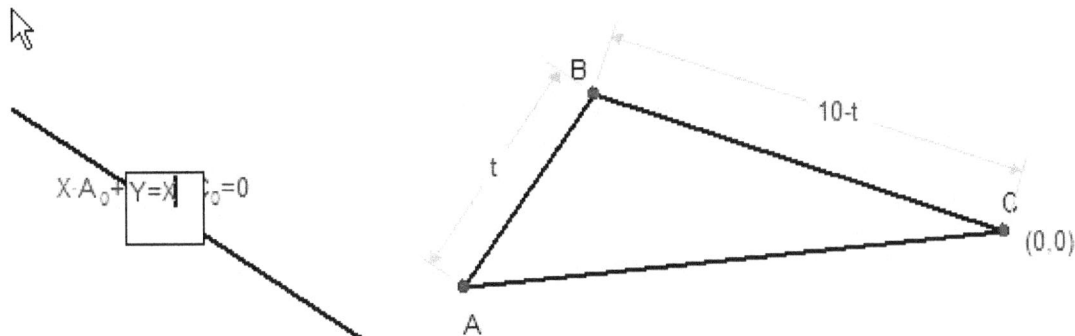

Notice how the line snaps into position after we press enter.

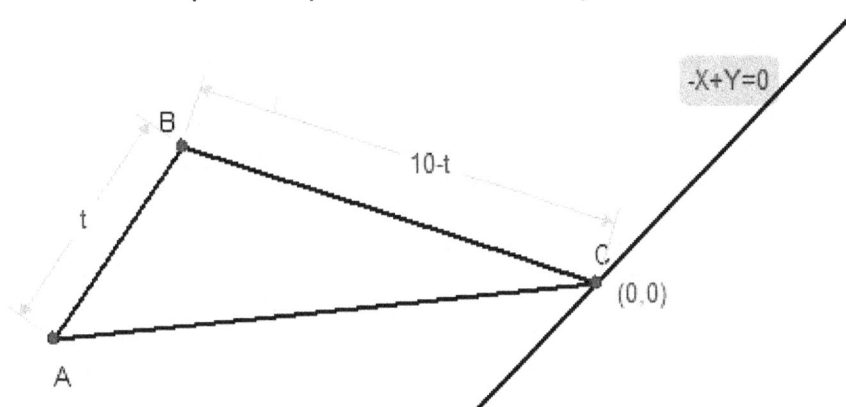

Constraining a Point Proportionally Along a Curve

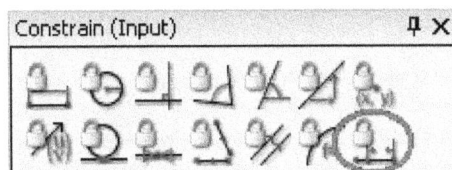

A point proportion t along a curve is defined variously for different types of curves as follows:

- For a **Line segment** AB, it defines the point $(1-t)*A + t*B$

- For a **Circle** it defines the point on the circle which subtends angle t at the center.

- For a **Locus** or envelope, it defines the point at parameter value t.

- For general **Cartesian** functions, it defines the x value of the point on the function.

- For **Polar** functions, it defines the point on the function which subtends angle t.

- For general **Parametric** functions, it defines the point at parameter value t.

- For an **Ellipse** of the form $X^2/a^2 + Y^2/b^2 = 1$ it defines the point $(a\cos(t), b\sin(t))$.

- For a **Parabola** of the form $Y=X^2/4a$ it defines the point $(2at, at^2)$

- For a **Hyperbola** of the form $X^2/a^2 - Y^2/b^2 = 1$ it defines the point $(a/\cos(t), (b\sin(t))/\cos(t))$.

1. Select a point and one of the curves mentioned above.

2. Click the **Point Proportional** icon from the **Constrain** toolbox, or select **Point Proportional** from the **Constrain** menu.

3. Enter the parameter or quantity in the data entry box.

For example, in the following diagram, D is defined proportion t along AB, and E is defined proportion t along BC. The curve is the locus of F as t varies between 0 and 1.

(4,4)

B

(0,0)

D

t

F   E

t

A

C

(5,-1)

$$7 \cdot X^2 + 32 \cdot Y + 7 \cdot Y^2 + X \cdot (-32 - 2 \cdot Y) = 0$$

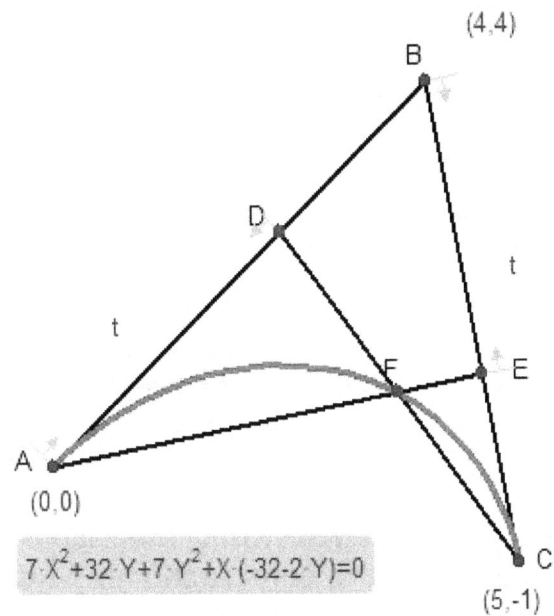

In the following example, the curve $Y = X^2$. Tangents are created at points with parameter values x_0 and x_1 on this curve. Point C is calculated for the current location of the tangents.

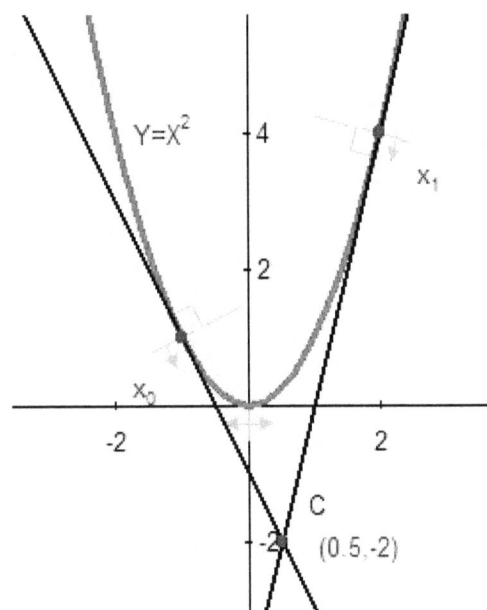

$Y = X^2$

4

x_1

2

x_0

-2   2

C

(0.5,-2)

Constructions

Creating Constructions

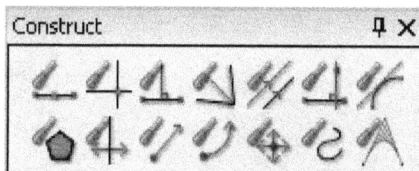

After sketching and constraining your drawing there are a whole set of constructions that can be applied to the geometry. First you must select the geometry elements which pertain to the construction. When you select the geometry the appropriate constructions will be highlighted.

The following table lists the **Constructions**, their icons, and which elements must be preselected to activate the constructions. Be careful when selecting geometry objects, if extra things are selected that are not related to the construction, the construction icons will remain inactive. This can happen by mistake, especially when using the selection box tool.

	Construction	**Preselected Objects**
	Midpoint	A line segment, vector, or polygon side.
	Intersection	Two of: a line, segment, vector, polygon side or conic. Conics are limited to intersecting only with lines, segments and vectors.
	Perpendicular Bisector	A line segment, vector, or polygon side.
	Angle Bisector	Two of: a line, segment, vector, or polygon side.
	Parallel	A point and one of: a line, segment, vector, or polygon side.
	Perpendicular	A point and one of: a line, segment, vector, or polygon side.
	Tangent	A circle or curve, and optionally, a point on the curve

	Polygon	Three or more connected line segments or points (vertices) to form a polygon
	Reflection	One or more objects
	Translation	One or more objects
	Rotation	One or more objects
	Dilation	One or more objects
	Locus	A point or line that will vary with a parameter
	Trace	One or more objects that will vary with a parameter

Midpoints of Line Segments

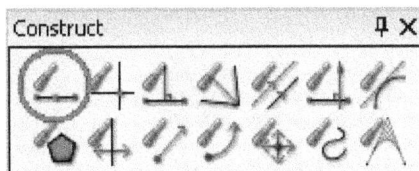

You can construct a midpoint on any line segment, vector, polygon side, or between two points by:

1. Select ⬚ two from the geometry types listed above.

2. Click the **Midpoint** tool , or select **Midpoint** from the **Construct** menu.

A point will appear in the middle of the selected line.

Intersections

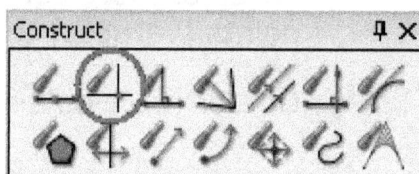

You can construct a point of intersection between any of the line types in your geometry; line, segment, vector, polygon side or circle. You can also construct intersections of circles. Conics are limited to intersections only with lines, segments or vectors.

1. Select two from the line types listed above.

2. Click the **Intersection** tool or select **Intersection** from the **Construct** menu.

A new point and label will appear at the intersection. If the lines are segments that do not intersect, a point will be created at the extension of the lines as with line segments AB and CD below.

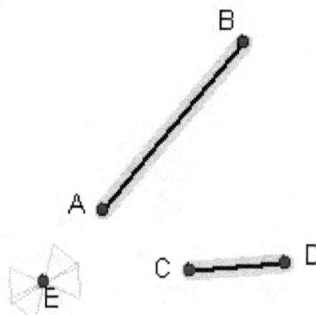

If the geometry will never intersect, the selected objects are moved to form the intersection. In the example below, the infinite line and circle become tangent at the newly created point, H.

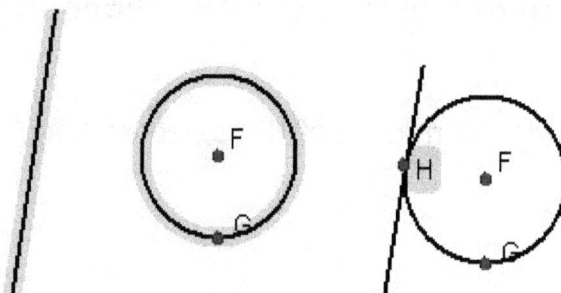

Perpendicular Bisector

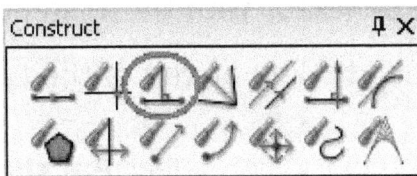

You can construct a perpendicular bisector on any line, segment, vector, or polygon side with these steps:

1. Select ⟦⟧ any of the line types listed above.

2. Click the **Perpendicular Bisector** tool ⟦⟧ or select **Perpendicular Bisector** from the **Construct** menu.

An infinite line will appear at right angles to the selected line.

Angle Bisector

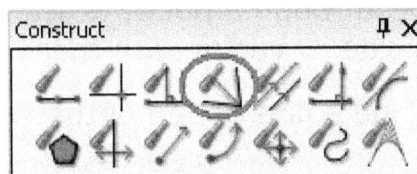

You can bisect the angle between any combination of line types; line, segment, vector, or polygon side with these steps:

1. Select ⟦⟧ two of any of the line types listed above.

2. Click the **Angle Bisector** tool ⟦⟧ or select **Angle Bisector** from the **Construct** menu.

An infinite line will appear between the two selected lines. You can use the Calculate / Angle tool to get the value of the bisected angle.

Parallel Constructions

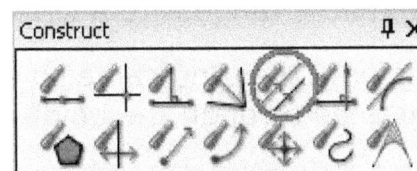

You can construct a line, through a point, and parallel to another line, segment, polygon side or vector with these steps:

1. Select ⬚ a point and a line of one of the types listed above.

2. Click the **Parallel** tool ⬚ or select **Parallel** from the **Construct** menu.

A line is constructed which is parallel to the selected line and passes through the selected point.

Perpendicular Constructions

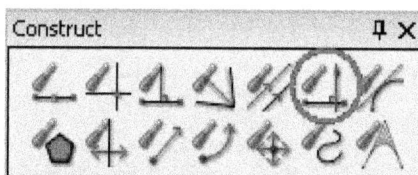

You can construct a line, through a point, which is perpendicular to another line, segment, polygon side or vector with these steps:

1. Select ⬚ a point and a line of one of the types listed above.

2. Click the **Perpendicular** tool ⬚ or select **Perpendicular** from the **Construct** menu.

A line is constructed which is perpendicular to the selected line and passes through the selected point.

Tangents

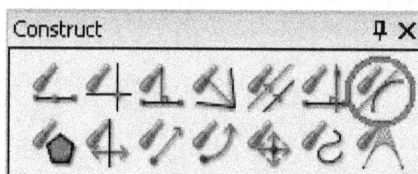

You can construct a line that is tangent to a circle or curve with these steps:

1. Select ⬚ the circle or curve. You can also select a point on the curve so that the tangent goes through the point on the curve.

2. Click the **Tangent** tool or select **Tangent** from the **Construct** menu.

A line tangent to the selected curve will appear at the point where you selected the circle or curve, or at the selected point.

Polygons

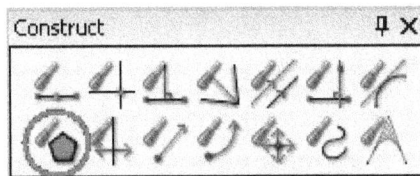

If you created a polygon with the line segment tool, or your polygon was not shaded for some reason, (*e.g.* the drawing of the sides was interrupted or out of order) you can make joined line segments into a polygon that can be selected with a single click using this construction.

1. Select the line segments that make up the polygon.

2. Click the **Polygon** tool in the **Construct** toolbox, or select **Polygon** from the **Construct** menu.

The polygon will be filled and you can now select the entire polygon with a single click.

Reflection

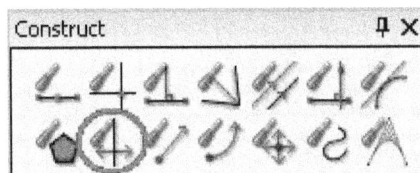

You can reflect any subset of your diagram about a line with these steps:

1. Select one or more geometry objects to reflect.

2. Click the **Reflection** tool in the **Construct** toolbox, or select **Reflection** from the **Construct** menu.

3. Either click the cursor to place the reflection line on the screen, adjust the angle and click again, or select an existing line as the reflection line.

A copy of your selected geometry will appear on the other side of the reflection line.

Notice all points on the reflected geometry are written as "prime", *i.e.* A becomes A'. If you reflect the geometry again, A' becomes A".

Translation

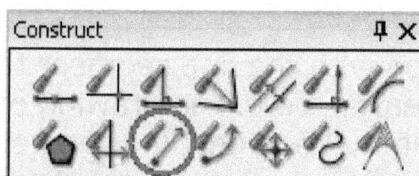

You can translate any subset of your diagram with a translation vector. Here are the steps:

1. Select the geometry to be translated.

2. Click the **Translation** tool in the **Construct** toolbox, or select **Translation** from the **Construct** menu.

3. Click the cursor to draw the end point of your translation vector and move the cursor to establish the length and angle of the translation. Click again to finish the vector.

The translated geometry appears. You can adjust the position of the translation by clicking and dragging the tip of the vector.

Notice all points on the translated geometry are written as "prime", *i.e.* A becomes A'. If you translate this geometry again, A' becomes A".

Rotation

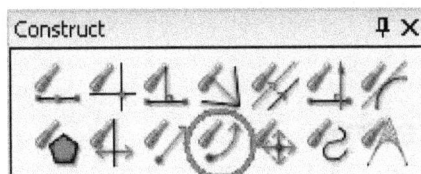

You can rotate any subset of your diagram about a point. Here are the steps:

1. Select the geometry to be rotated.

2. Click the **Rotation** tool in the **Construct** toolbox, or select **Rotation** from the **Construct** menu.

3. Click the screen to place your rotation point.

4. In the data entry box presented, enter the angle of rotation.

The rotation of the selected geometry appears.

Notice all points on the rotated geometry are written as "prime", *i.e.* A becomes A'. If you rotate this geometry again, A' becomes A".

Dilation

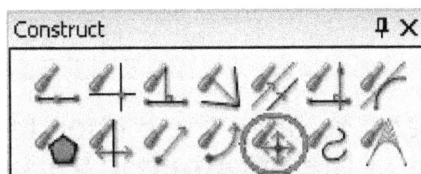

You can dilate any subset of your diagram from a point. Here are the steps:

1. Select the geometry to be dilated.

2. Click the **Dilation** tool in the **Construct** toolbox, or select **Dilation** from the **Construct** menu.

3. Click the cursor on your dilation point.

4. In the data entry box presented, enter the dilation factor.

The dilated geometry appears.

Notice all points on the dilated geometry are written as "prime", *i.e.* A becomes A'. If you dilate this geometry again, A' becomes A".

Locus of Points / Envelope

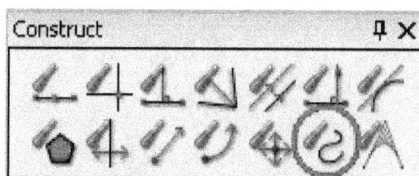

You can construct a locus of points or envelope from a selected point or line, by defining a range for some constraint in the drawing. Just follow these easy steps:

1. Select the point on the drawing that will form the locus (point B in the example below); select a line, line segment, or vector to form an envelope.

2. When you click the **Locus** icon , the **Edit Locus** dialog pops up.

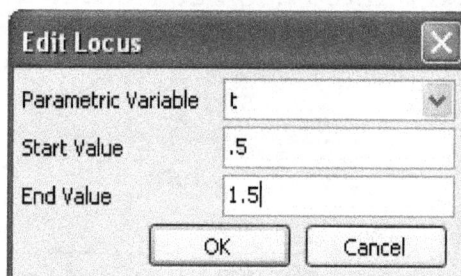

3. You need a parameter to drive the motion to create the locus. Click the arrow key to the right of the **Parametric Variable** window to select from a list of all variables in the drawing. (If you entered the needed constraint in real terms, Cancel the trace and change the constraint to a variable by double-clicking it in the drawing window.)

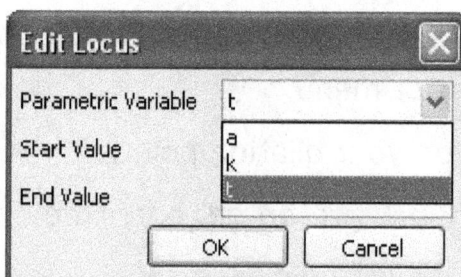

4. Simply fill in the values for the appropriate variable and click the Ok button.

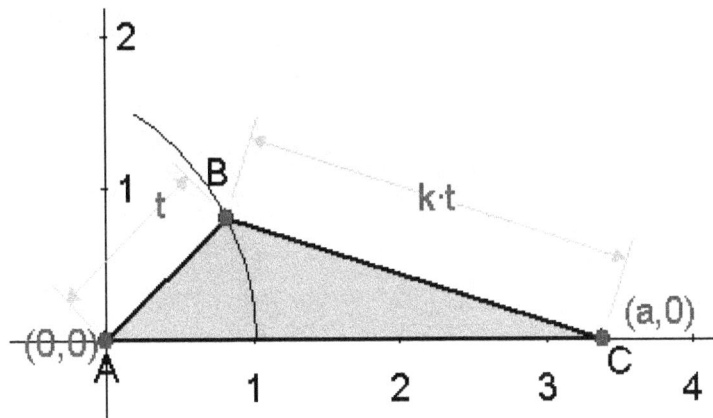

! Note: The locus only works if the figure's position is fixed i.e. a point in a triangle will not work as a locus unless the location of the other two points are fixed with coordinate constraints.

To adjust the range of the locus, double-click the locus to edit the dialog.

The following example shows an envelope of the line DE. We use the Point proportional along curve constraint and the parameter t to position the points D and E (D is (1-t) along line AC and E is t along line AB). In the **Edit Locus** dialog, we create the envelope from parameter t as it ranges from 0 to 1.

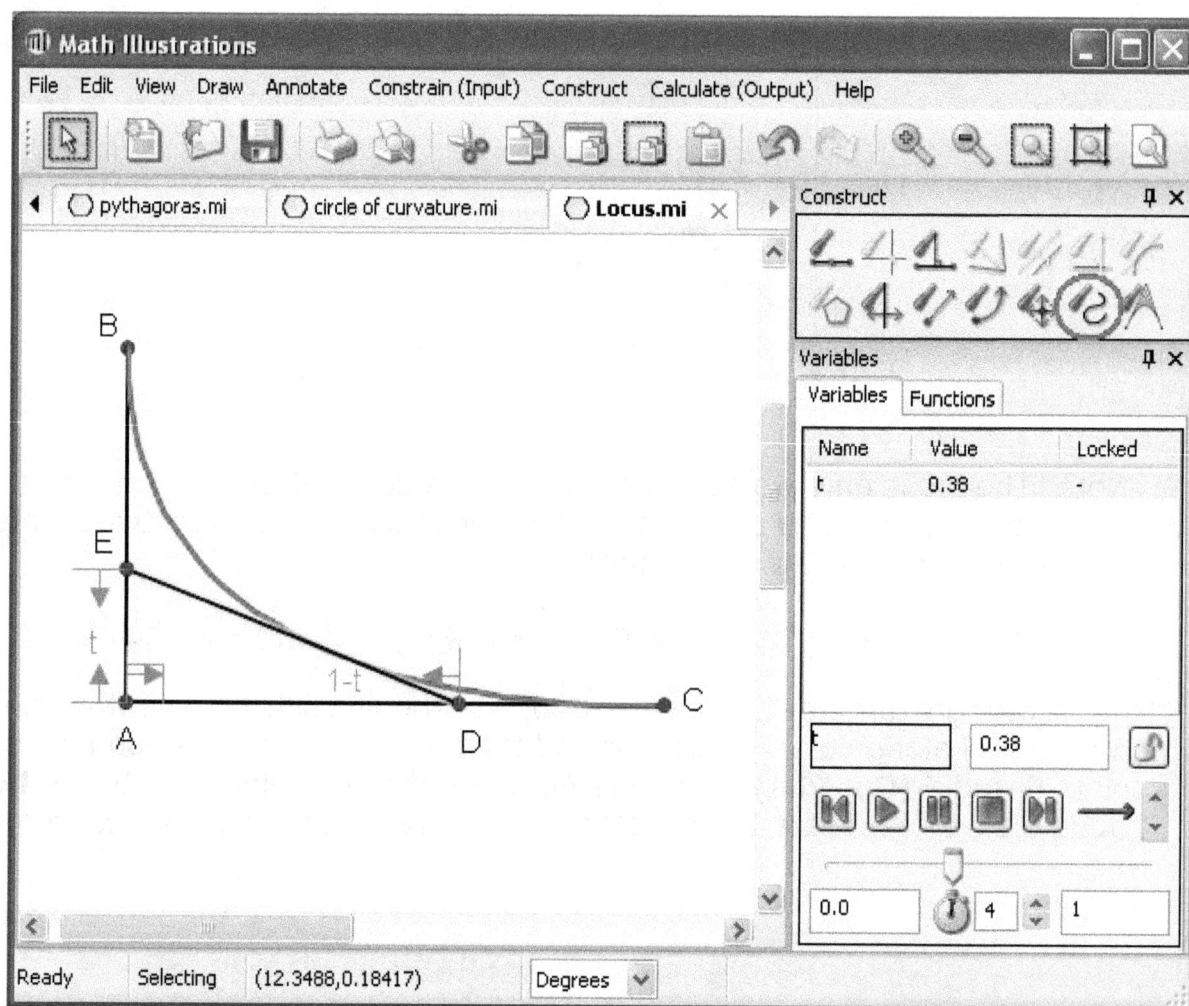

Trace

You can **Trace** the movement of one or a group of drawing objects. You can create string art drawings and see how an envelope curve is formed. Here are the steps:

1. Select the drawing objects to be traced.

2. Click the **Trace** tool and the **Edit Trace** dialog pops up.

3. You need a parameter to drive the motion of the trace. Click the arrow key to the right of the **Parametric Variable** window to select from a list of all variables in the drawing. (If you entered the needed constraint in real terms, Cancel the trace and change your constraint to a variable by double-clicking it in the drawing window.)

Edit Trace	☒		Edit Trace	☒
Parametric Variable	t ⌄		Parametric Variable	t ⌄
Start Value	0.14357171676		Start Value	a
End Value	5			k
			End Value	t
Count	20		Count	20
OK	Cancel		OK	Cancel

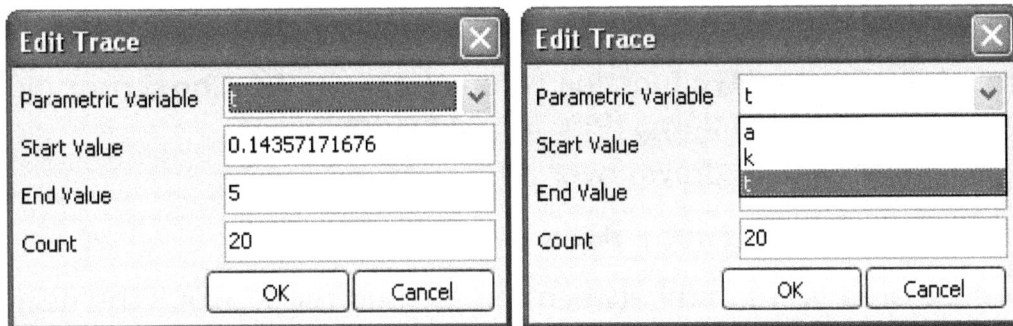

4. Simply fill in the values for the appropriate variable and click the OK button.

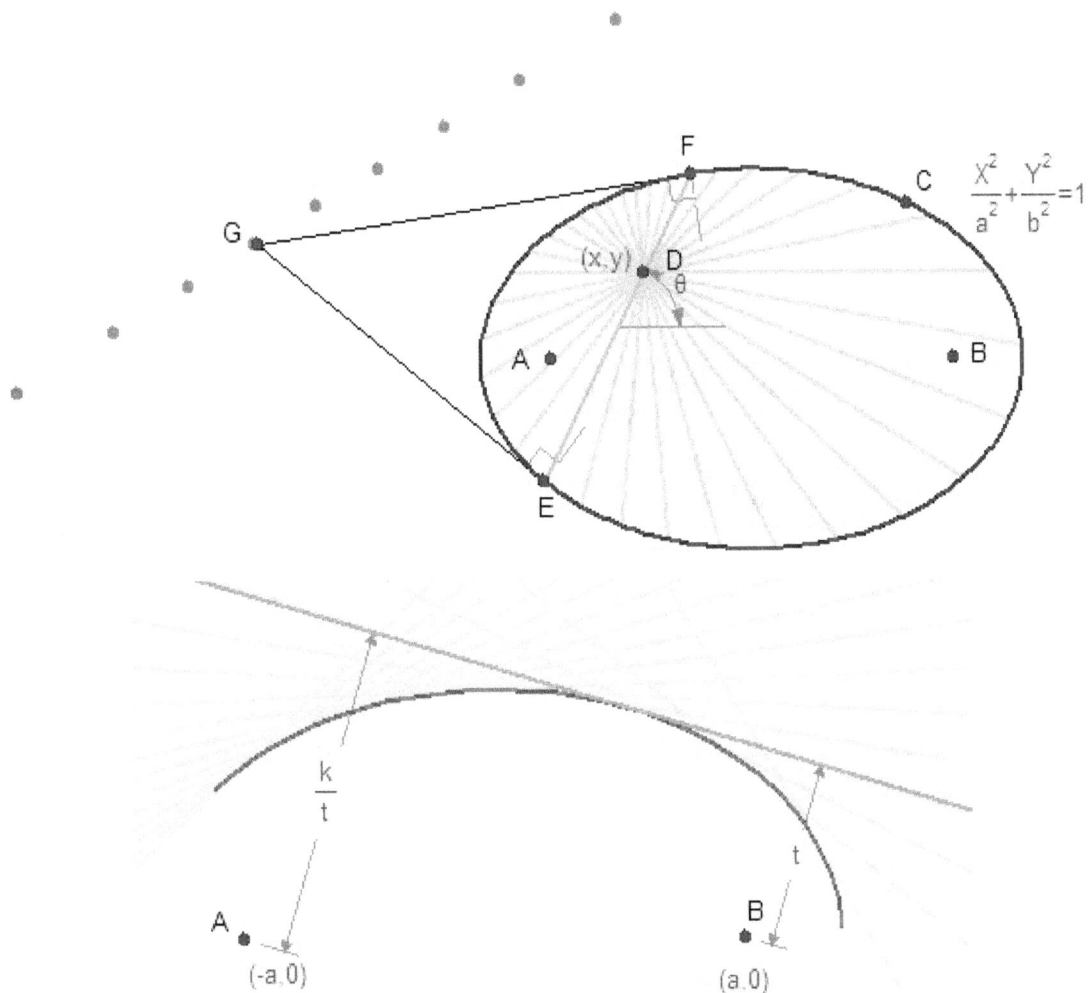

$$\frac{x^2}{a^2}+\frac{y^2}{b^2}=1$$

To adjust the range or number of traces, double-click one of the traces to edit the dialog.

Area Under the Arc

The **Area Under Arc** function is found only in the **Construct** menu at the top of the main window. Here are the steps:

1. Select an arc drawn over a function.

2. Select **Construct / Area Under Arc**.

A filled area is created between the arc and the X axis. The necessary lines and points containing the area are automatically added.

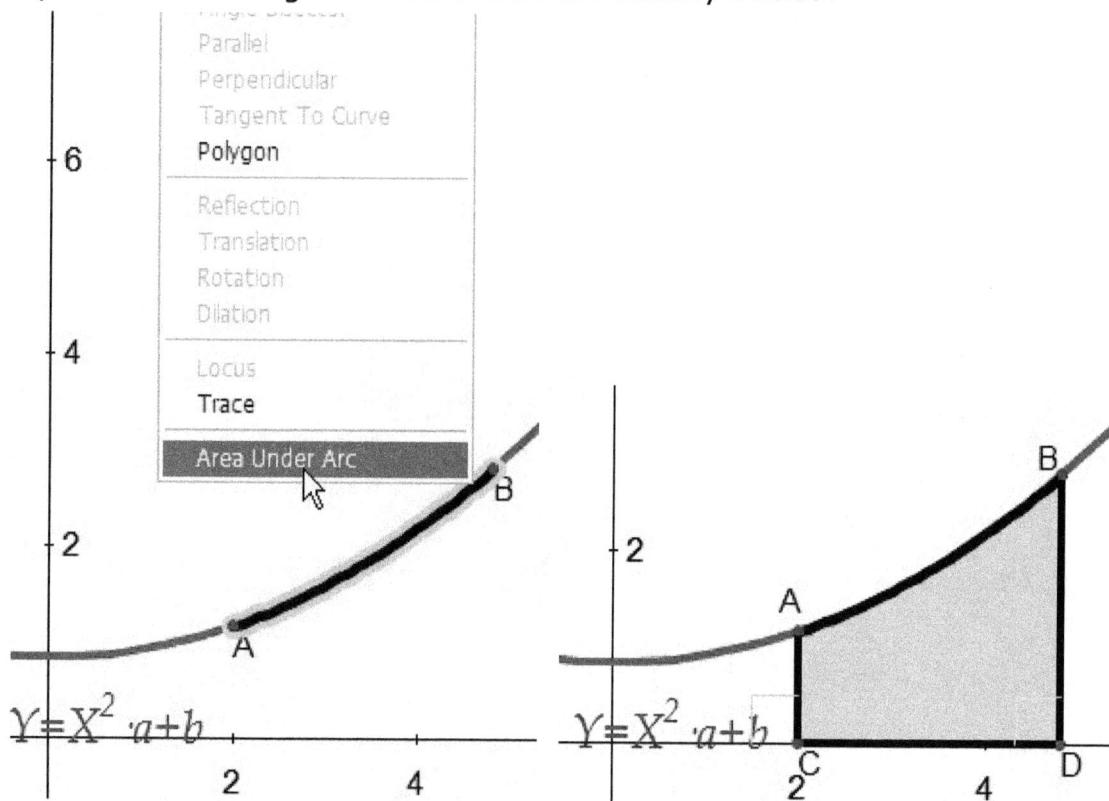

Drag the handles of the original function and the area under the curve changes accordingly.

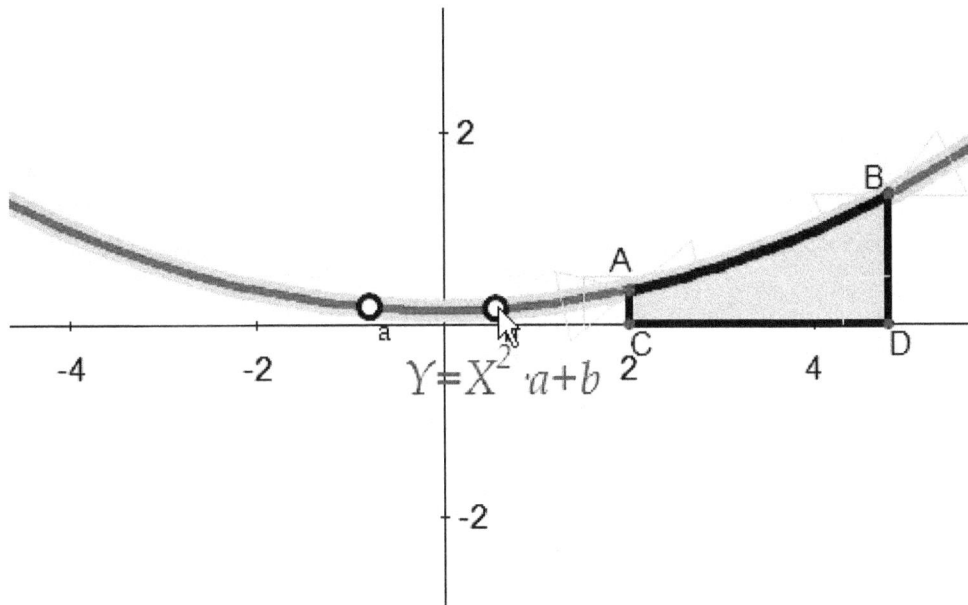

$$Y = X^2 \cdot a + b$$

Note: this only works for arcs drawn on functions. For arcs drawn on conics, create the sides with the **Draw / Line Segment**, select all sides and use the **Construct / Polygon** tool. See also: instructions for creating curvilinear polygons.

Annotations

Applying Annotations

Annotations allow you to add constraint information to your drawing which isn't needed for it's construction. These might be constraints that cause the geometry to be over constrained, but you might want to give the viewer some additional information. This feature can be very helpful for making up texts or worksheets.

Annotate tools are similar to **Text** in that they have no influence on the geometry engine, even though they are placed exactly like the **Constrain** tools.

Since annotations may look identical to constraints, use the icon, , to **Distinguish Constraints / Annotations**. The icon is a toggle; to turn off the marks, click it again.

The **Symbols** and **Annotation Symbols** toolboxes are both available to the **Annotate** tools.

Here are the **Annotate** tools and the object(s) to preselect:

	Annotation	Preselected Object(s)
	Distance / Length	A line segment, vector, or polygon side, or a point and one of these line types (perpendicular distance), or two points.
	Radius	Circle
	Perpendicular	Two of any line, segment, vector, or polygon side.
	Angle	Two of any line, segment, vector, or polygon side.
	Direction	A line, segment, vector, or polygon side.
	Slope	A line, segment, vector, or polygon side.
	Coordinates	Point
	Coefficients	Vector
	Congruent	A line, segment, vector, or polygon side.
	Congruent Angle	Two of any lines, segments, vectors, or polygon sides
	Parallel	A line, segment, vector, or polygon side.
	Expression	[none]

Distance / Length Annotation

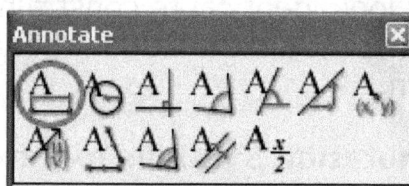

Length annotations may be applied to any line segment, polygon side or vector. Distance annotations are available between two points, or the perpendicular distance between a point and a line, segment, polygon side, or vector. Use these steps:

1. Select the line segment, or point and line, or pair of points as described above. When you make your selection, the drawing objects will be highlighted and the **Distance / Length** icon will light up .

2. Click the icon

3. Enter the distance information. There are no format restrictions. You can use Symbols and Annotation Symbols in the entry. Press enter when you're done.

You can click the annotation and drag it to adjust its placement in the drawing.

Radius Annotation

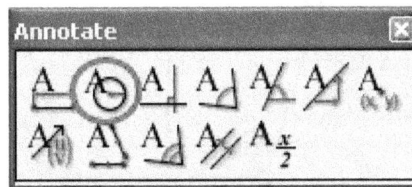

To annotate a circle's radius use these steps:

1. Select the circle. The **Annotate / Radius** icon will light up and the circle will be highlighted.

2. Click the icon

3. Enter the radius information. There are no format restrictions. You can use Symbols and Annotation Symbols in the entry. Press enter when you're done.

You can click the annotation and drag it to adjust its placement in the drawing.

Perpendicular Annotation

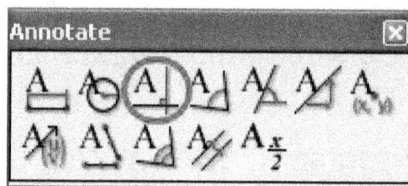

The perpendicular annotation inserts the perpendicular mark between any two of these line types: line, segment, polygon side or vector. Use these steps:

1. Select [cursor icon] two of any line, segment, vector, or polygon side. The **Annotate / Perpendicular** icon [icon] will light up when both lines are highlighted.
2. Click the icon.

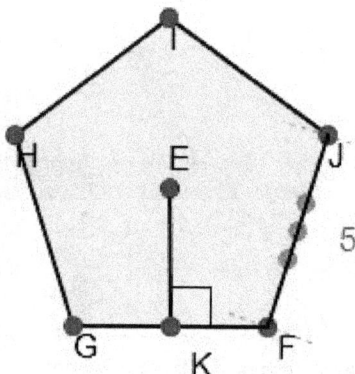

Note: Placing this annotation between two lines does <u>not</u> change the relative position of the lines or prevent the lines from changing their relative position as it does with the **Constrain** tool of the same name.

This annotation is very useful if you are displaying a 3-d object. In this example we imported the Triangular Prism from the **Solids** folder in the Figure Gallery.

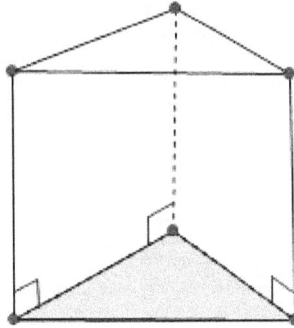

Angle Annotation

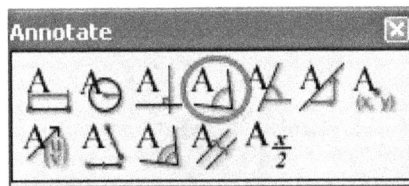

To annotate an angle use these steps:

1. Select two of any line, segment, vector, or polygon side. The

    Annotate / Angle icon will light up and the lines will be
    highlighted.
2. Click the icon
3. Enter the angle information. There are no format restrictions. You can use Symbols and Annotation Symbols in the entry. Press enter when you're done.

You can click the annotation and drag it to adjust its placement in the drawing.

Which Side to Annotate?

Sometimes when identifying angles, the annotation falls on the wrong one. In the example below, we wanted BDC, not BDA. Just click the cursor over the annotation arrow and drag it to the other side, then release the mouse button - done!

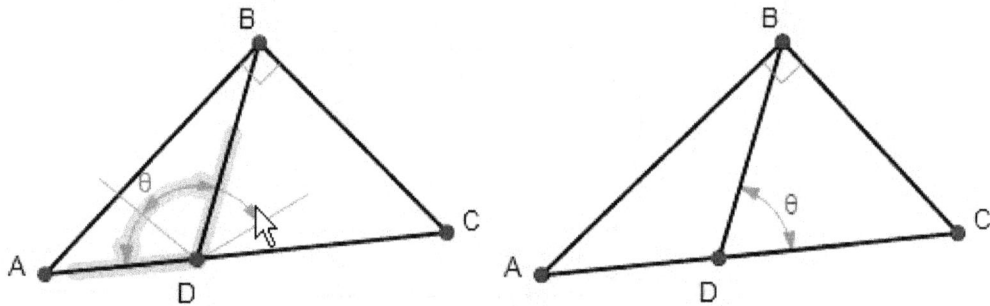

To get the reflex angle, hold the *Ctrl* key while dragging the angle symbol.

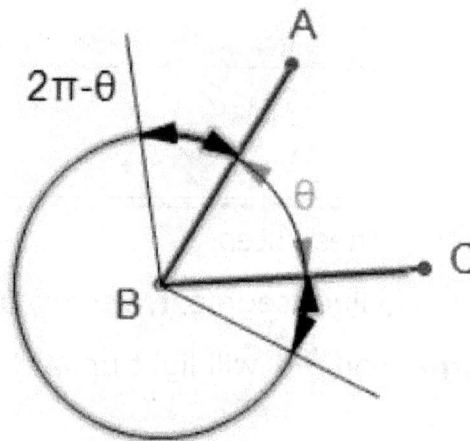

Direction Annotation

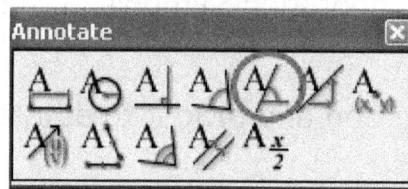

To annotate a line's direction use these steps:

1. Select ⬚ a line, segment, vector, or polygon side. The **Annotate / Direction** icon ⟁ will light up and the line will be highlighted.

2. Click the icon

3. Enter the direction information. There are no format restrictions. You can use Symbols and Annotation Symbols in the entry. Press enter when you're done.

You can click the annotation and drag it to adjust its placement in the drawing.

Slope Annotation

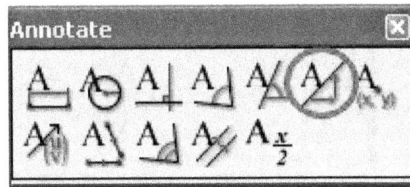

To annotate a line's slope use these steps:

1. Select [cursor] a line, segment, vector, or polygon side. The **Annotate / Slope** icon [icon] will light up and the line will be highlighted.
2. Click the icon
3. Enter the slope information. There are no format restrictions. You can use Symbols and Annotation Symbols in the entry. Press enter when you're done.

You can click the annotation and drag it to adjust its placement in the drawing.

Coordinate Annotation

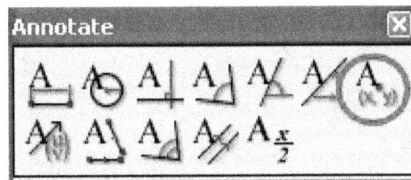

To annotate an point's coordinates use these steps:

1. Select [cursor] point. The **Annotate / Coordinate** icon [icon] will light up and the point will be highlighted.
2. Click the icon
3. Enter the coordinate information. There are no format restrictions. You can use Symbols and Annotation Symbols in the entry. Press enter when you're done.

You can click the annotation and drag it to adjust its placement in the drawing.

Coefficients Annotation

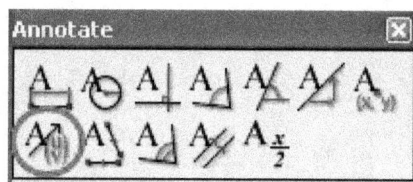

To annotate a vector's coefficients use these steps:

1. Select ⬚ vector. The **Annotate / Coefficients** icon ⬚ will light up and the vector will be highlighted.

2. Click the icon

3. Enter the vector's coefficients. There are no format restrictions. You can use Symbols and Annotation Symbols in the entry. Press enter when you're done.

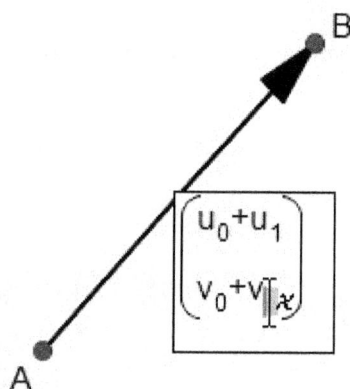

You can click the annotation and drag it to adjust its placement in the drawing.

Congruent Annotation

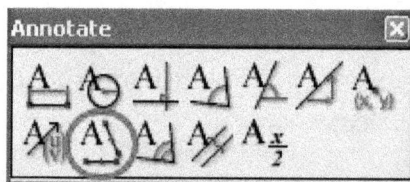

Place a congruent mark on any of the linear drawing elements: lines, segments, vectors, or polygon sides. Use these steps:

1. Select ⬚ any line, segment, vector, or polygon side. The

   Annotate / Congruent icon ⬚ will light up when both lines are highlighted.

2. Click the icon.

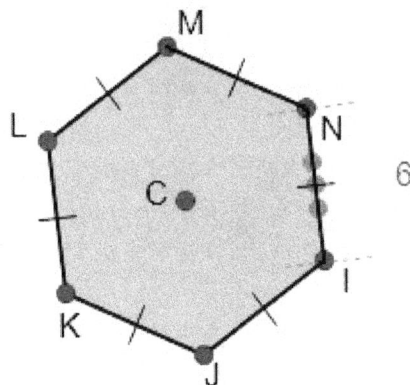

Note: Placing this annotation on lines does <u>not</u> change the relative lengths of the lines or keep them the same length as it does with the **Constrain** tool of the same name.

Congruent Angle Annotation

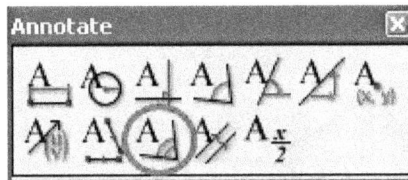

Place a congruent mark between pairs of linear drawing elements: lines, segments, vectors, or polygon sides. Use these steps:

1. Select ⬚ two lines, segments, vectors, or polygon sides. The

   Annotate / Congruent icon will light up when both lines are highlighted.

2. Click the icon.

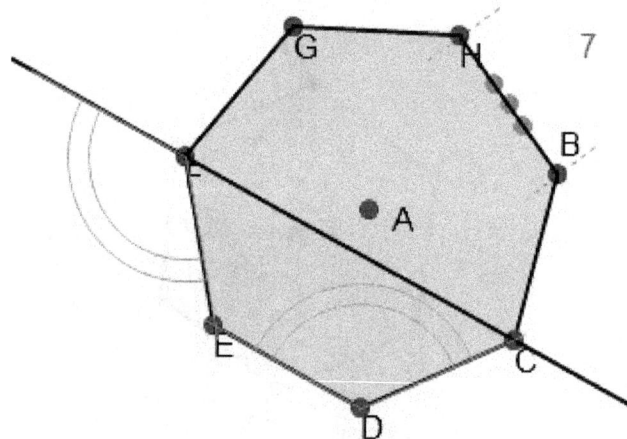

You can change the arc count with these steps:

1. Select the annotation.

2. Right click to invoke the selection Context menu.

3. Select **Tic/Arc Count** from the menu and click the desired number.

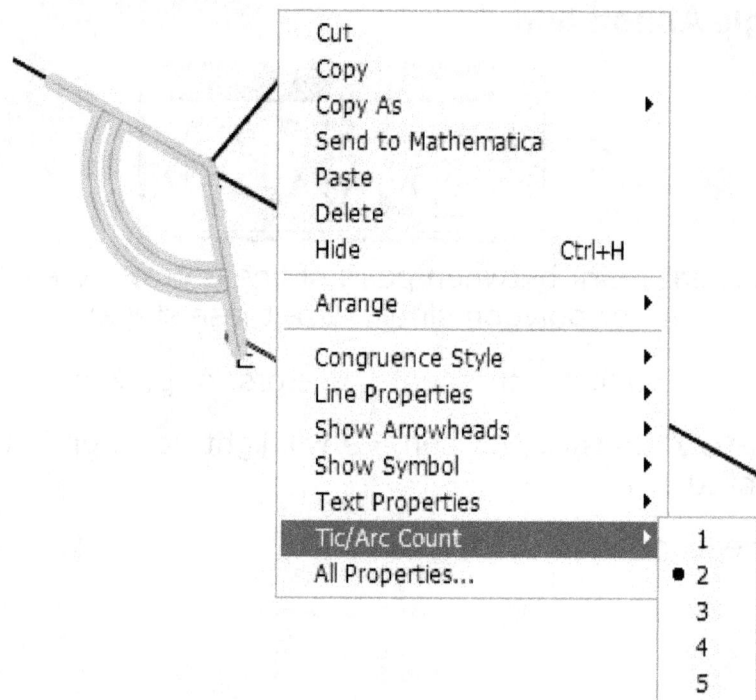

Parallel Annotation

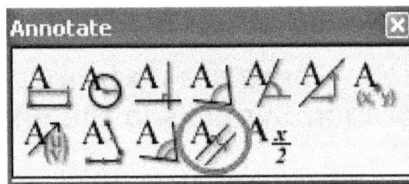

Place a parallel mark on any of the linear drawing elements: lines, segments, vectors, or polygon sides. Use these steps:

1. Select ⌞⬚⌟ from one to three of any line, segment, vector, or polygon side. The **Annotate / Perpendicular** icon will light up when both lines are highlighted.

2. Click the icon.

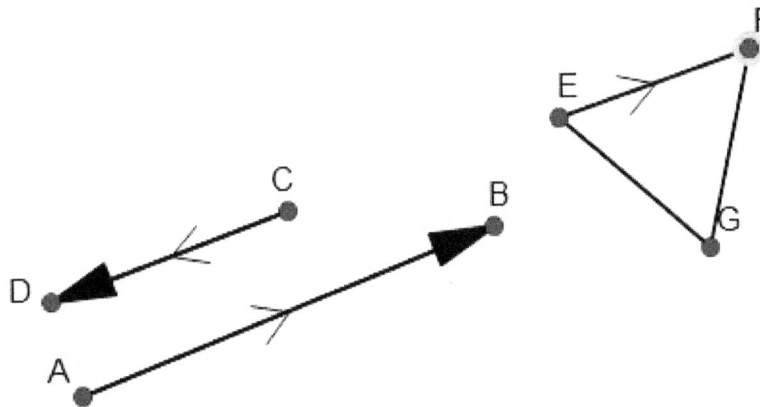

Note: Placing this annotation on lines does **not** change the relative position of the lines or prevent the lines from changing their relative position as it does with the **Constrain** tool of the same name.

Expression Annotation

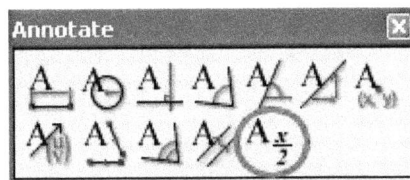

The **Annotate / Expression** command is useful for placing a line of mathematics anywhere in your drawing.

1. Click the icon $A_{\frac{x}{2}}$.

2. Move the cursor to the position where you want to place the expression and click.

3. Enter the expression. There are no format restrictions. You can use Symbols and Annotation Symbols in the entry. Press enter when you're done.

You can click the annotation and drag it to adjust its placement in the drawing.

Calculations

Calculating the Output

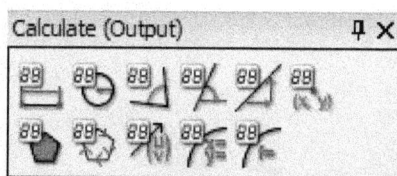

Math Illustrations will make calculations in the geometry based on any constraints or constructions you have specified, or just from the sketch. Calculations are output only in **Real** terms. If you are interested in **Symbolic** manipulations check out *Geometry Expressions.* You can download a demo at: www.GeometryExpressions.com.

If you haven't supplied all of the necessary input constraints for an exact answer, the system bases the output value on the sketch.

The table below lists all the available calculations and geometry elements which must be preselected. Be careful when selecting geometry objects, if extra things are selected that are not related to the calculation (like other calculations) the calculations will remain inactive. This can happen by mistake, especially when using the selection rectangle.

	Calculation	Preselected Object(s)
	Distance / Length	A line segment, vector, or polygon side, or a point and one of these line types (perpendicular distance), or two points.
	Radius	Circle
	Angle	Two of any line, segment, vector, or polygon side.
	Direction	A line, segment, vector, or polygon side.
	Slope	A line, segment, vector, or polygon side.
	Coordinates	Point
	Area	Circle or polygon
	Perimeter	Circle or polygon
	Coefficients	Vector
	Parametric Equation	A line, segment, vector, polygon side, circle, or a constructed locus.
	Implicit Equation	A line, segment, vector, polygon side, circle, or a constructed locus.

Assigning Properties to the Output

You can change the **Display Properties** of the output, including Line Color, Line Style, Font, Line Equation Style, Decimal Digits, and whether or not a name is shown with the output.

After you generate an output expression:

1. click the output

2. right click the mouse and select **All Properties** from the **Selection**

context menu or from the **Edit** menu

3. click one of the rows

4. click the drop-down arrow or to select a different property.

These properties are also included in the default settings under: **Edit / Preferences / Math**. Note: changing these settings does <u>not</u> change properties of elements already on the drawing. Change existing elements by selecting them and using the **Properties** command as described above.

<u>**Show Name**</u> - is a term assigned by the system to the output. This name is z_n where n is the sequential number of the output .

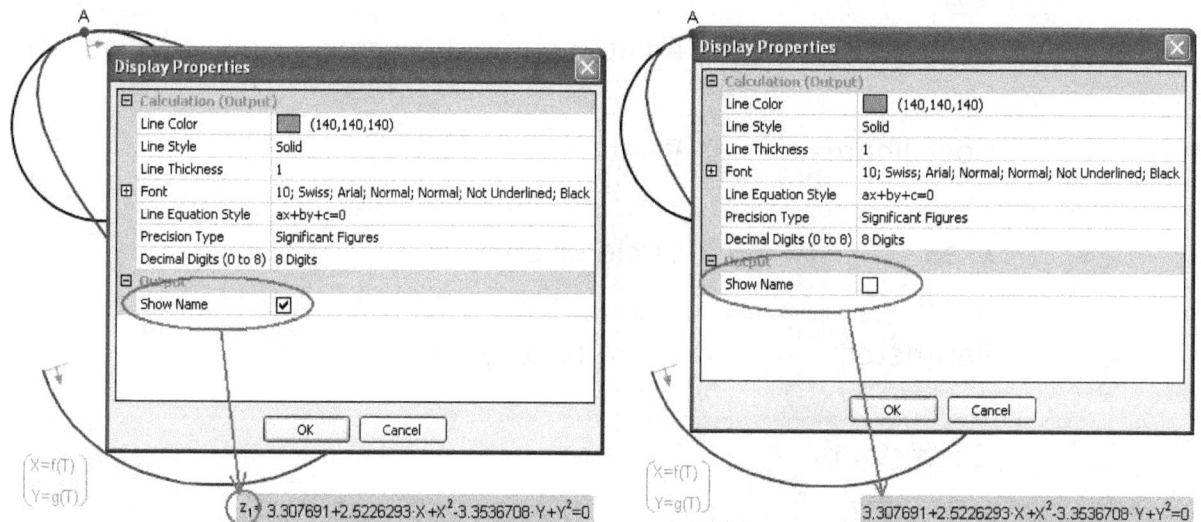

These names are a handy shortcut for using in the **Draw / Expression** tool.

Distance / Length Calculation

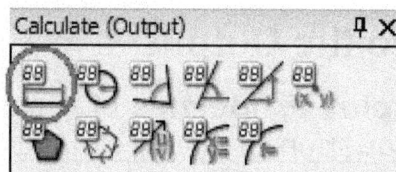

Length calculations may be obtained for any line segment, arc, polygon side or vector. Distance calculations are available between two points, or the perpendicular distance between a point and a line, segment, polygon side, or vector. Use these steps:

1. Select ⬚ the line segment, arc, or point and line, or pair of points as described above.

2. Click the **Distance / Length** tool in the **Calculate** toolbox or select **Distance / Length** from the **Calculate** menu.

Math Illustrations displays the length, using any relevant parameters you may have specified.

Radius Calculation

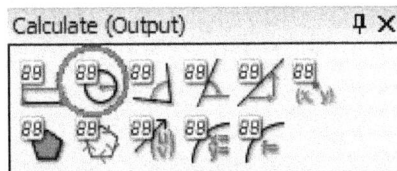

Math Illustrations will calculate the radius of any circle. Use these steps to find the radius:

1. Select ⬚ a circle.

2. Click the **Radius** tool in the **Calculate** toolbox or select **Radius** from the **Calculate** menu.

The radius value appears in the diagram.

Angle Calculation

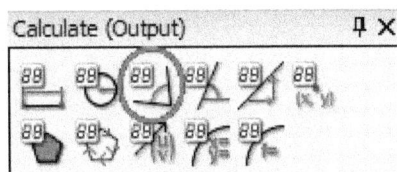

Math Illustrations will calculate any angle between lines in the geometry. Use these steps to find the angle:

1. Select ⬚ two line types - any line, segment, vector, or polygon side.

2. Click the **Angle** tool in the **Calculate** toolbox or select **Angle** from the **Calculate** menu.

The angle value in degrees or radians, depending on your **Preferences**, appears in the diagram.

You can obtain the angle's supplement by dragging the angle symbol.
To get the reflex angle, hold the *Ctrl* key while dragging the angle symbol.

Supplementary and Reflex Angles

If it's unclear whether a calculation is requested for the angle or its supplement, you can drag the angle symbol to the correct position.

Here are some examples of playing around with supplementary angles, both inputs (shown in blue) and outputs (shown in red):

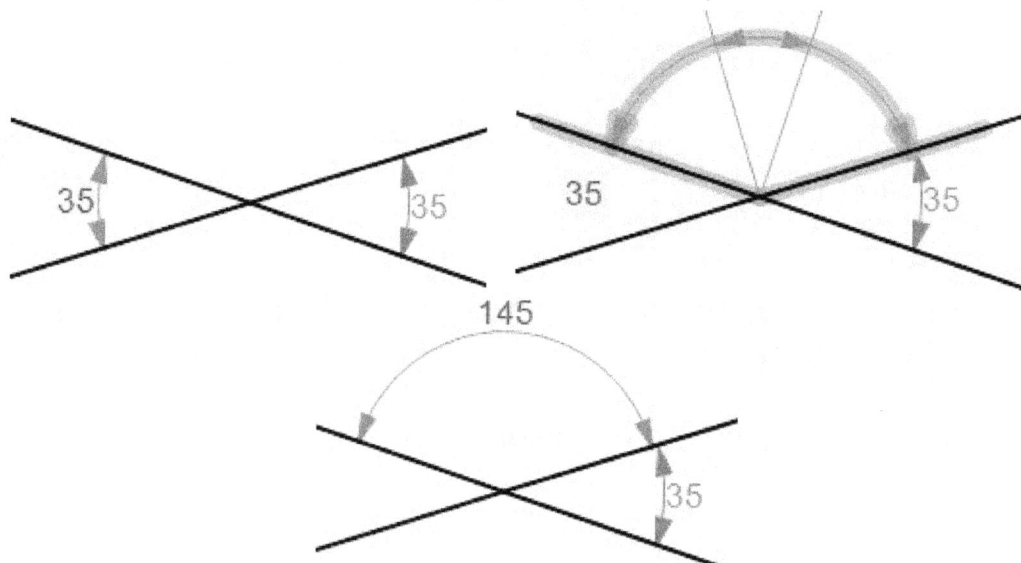

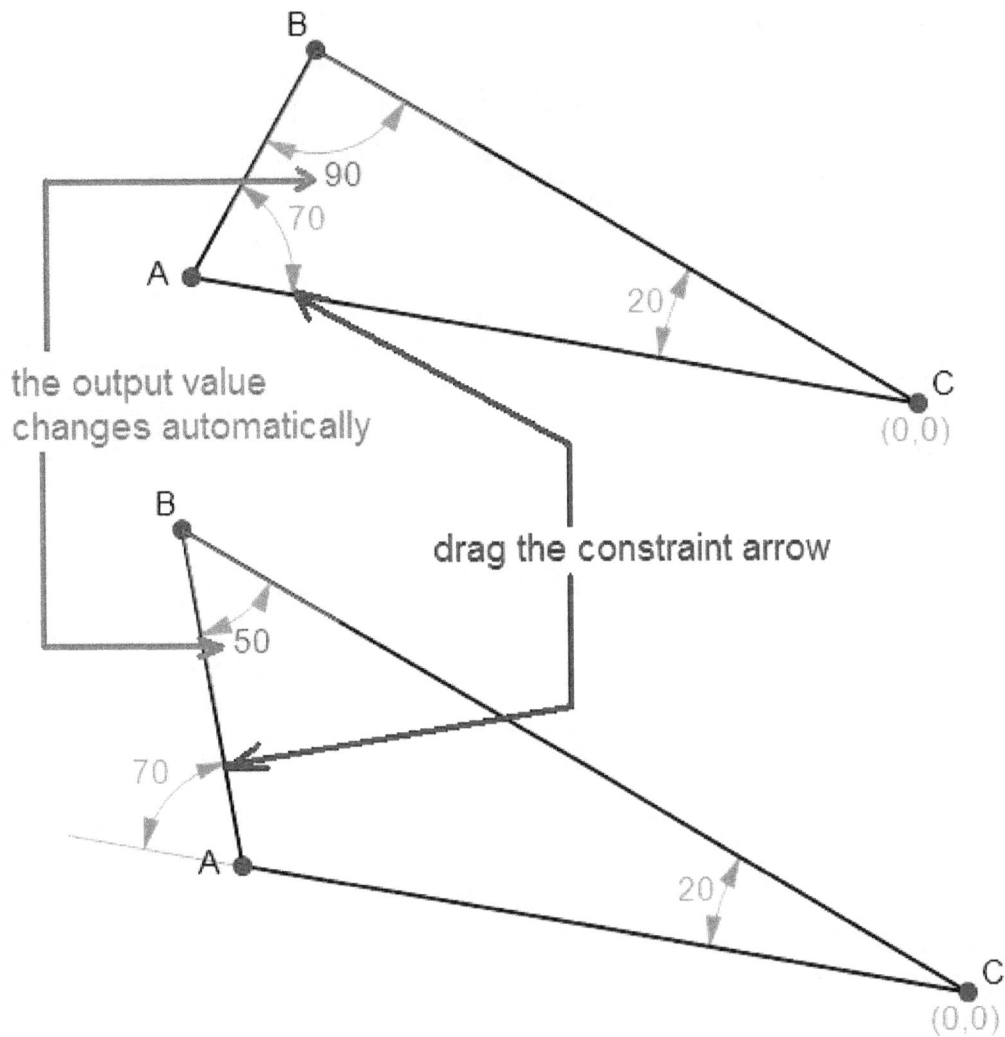

B

90
70

A

20

C
(0,0)

the output value
changes automatically

B

drag the constraint arrow

50

70

A

20

C
(0,0)

To get the reflex angle, hold the *Ctrl* key while dragging the angle symbol.

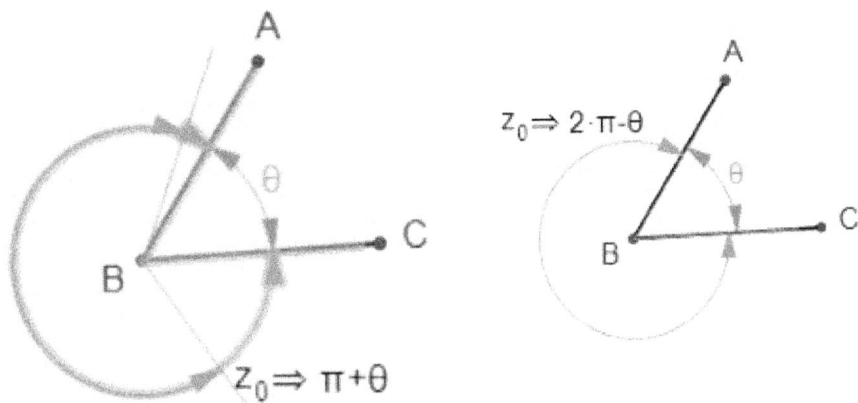

A

θ

B

C

$z_0 \Rightarrow \pi+\theta$

$z_0 \Rightarrow 2\cdot\pi-\theta$

A

θ

B

C

Direction Calculation

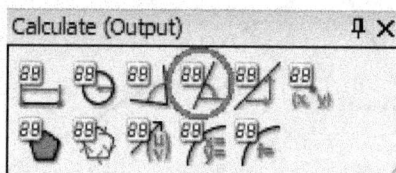

Math Illustrations will calculate the direction of lines, segments, polygon sides, or vectors with these steps:

1. Select [cursor] a line type.

2. Click the **Direction** tool in the **Calculate** toolbox or select **Direction** from the **Calculate** menu.

The direction measurement appears in degrees or radians.

Slope Calculation

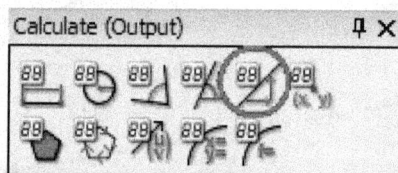

Math Illustrations will calculate the slope of lines, segments, polygon sides, or vectors with these steps:

1. Select [cursor] a line type.

2. Click the **Slope** tool in the **Calculate** toolbox or select **Slope** from the **Calculate** menu.

The equation for the slope appears in the diagram.

Calculate Coordinates

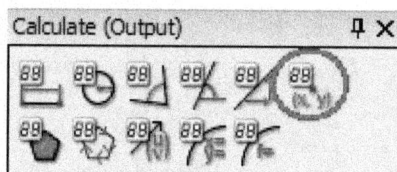

You can calculate the coordinates of any point in your diagram with these steps:

1. Select ⬚ a point.

2. Click the **Coordinates** tool in the **Calculate** toolbox or select **Coordinates** from the **Calculate** menu.

The coordinates appear by the point.

Area Calculation

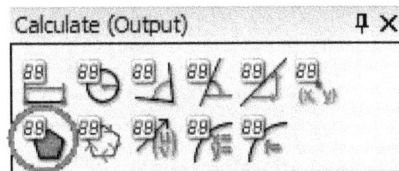

You can obtain the area of any polygon or circle in your diagram.

Note: If your polygon is not filled it is just a group of line segments. To convert them to a polygon, use the Polygon Construction tool, then proceed with these steps:

1. Select ⬚ a circle or polygon.

2. Click the **Area** tool in the **Calculate** toolbox or select **Area** from the **Calculate** menu.

The area is displayed.

Perimeter Calculation

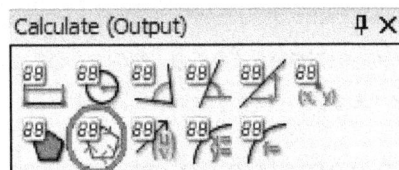

You can obtain the perimeter of any polygon or circle in your diagram.

Note: If your polygon is not filled it is just a group of line segments. To convert them to a polygon, use the Polygon Construction tool, then proceed with these steps:

1. Select a circle or polygon.

2. Click the **Perimeter** tool in the **Calculate** toolbox or select **Perimeter** from the **Calculate** menu.

The perimeter is displayed.

Calculate Coefficients

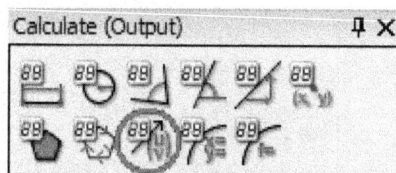

Use this tool to calculate the coefficients of a vector in the diagram with these steps:

1. Select a vector.

2. Click the **Coefficients** tool in the **Calculate** toolbox or select **Coefficients** from the **Calculate** menu.

The coefficients appear by the vector.

Calculating Parametric Equations

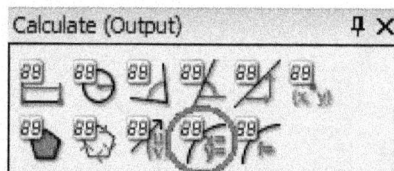

Computes parametric equations for a locus or envelope, based on the parameter defining the curve.

You can also calculate parametric equations for a circle or line.

Use these steps:

1. Select any geometry object described above.

2. Click the **Parametric Equation** tool in the **Calculate** toolbox or select **Parametric Equation** from the **Calculate** menu.

The equations for x and y appear by the geometry.

Calculating Implicit Equation

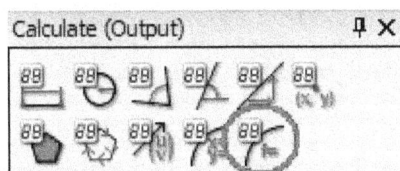

Calculates the implicit equation for the selected circle or a line.

Math Illustrations will also attempt to calculate the equation of a locus or envelope curve.

1. Select [⬚] any geometry object described above.

2. Click the **Implicit Equation** tool in the **Calculate** toolbox or select **Implicit Equation** from the **Calculate** menu.

The equation appears by the geometry.

Symbols

Using Symbols

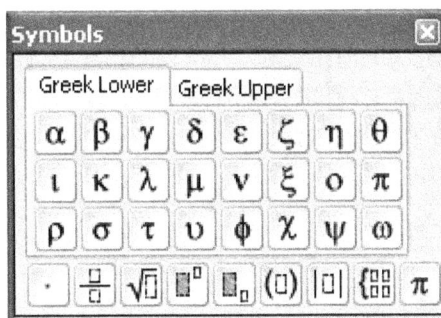

The **Symbols** toolbox lets you easily insert Greek letters into your expressions, constraints and annotations. Click the tab to choose from lower case or upper case Greek letters.

The bottom row of buttons in the toolbox lets you insert commonly used math operations. You can either use the icons, or your type them from your keyboard:

Symbol Icon	Function Call / Reserved Word
√□	sqrt(value)
|□|	abs(value)
{⊞}	piecewise({*expression1, domain1*},{*expression2, domain2*}...,{*last expression, otherwise*})
π	pi

Inserting Greek Letters

To insert Greek letters into any variable name or expression, click the appropriate tab, **Greek Upper** (upper case letters) or **Greek Lower** (lower case letters), and click the letters to be inserted into the data entry box.

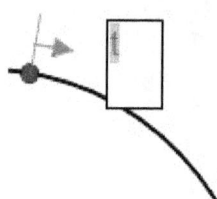

entry highlighted

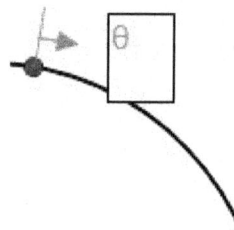

click the symbol
press the enter key

If your **Symbols** toolbox is hidden, you might want to just type the name of the Geek letter into your expression. The symbol will be inserted after you press *enter*. To get an uppercase Greek symbol, capitalize the first letter of it's name.

Square Root Editing Tool

√□ You can enter square roots in these ways:

- From the data entry box, enter the expression you want inside the square root, highlight the terms, and click the **Square Root** button.

- From the data entry box, click the **Square Root** button, then highlight the 0 and type the terms.

- Use the *sqrt()* function in the data entry box.

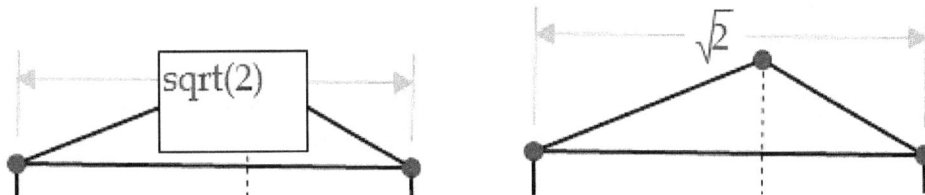

Multiplication & Division Editing Tools

The **Multiplication** button inserts a multiplication symbol into the expression.

The **Division** button makes expressions easier to enter and read.

- From the data entry box, enter the numerator of the expression, highlight it, and then click **Division**.

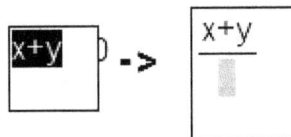

The cursor is then positioned in the denominator.

- If you click the **Division** button first, be sure to place the cursor in the appropriate place before typing the expression.

Subscript / Superscript Editor

You can enter superscripts or subscripts for variables in one of these ways:

- From the data entry box, enter the expression you want sub/superscripted, highlight the terms and click the **Subscript** or **Superscript** button.

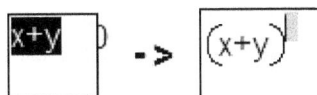

- From the data entry box, click the **Sub/Superscript** button and type the values into the grey boxes.

Note: Make sure the cursor is positioned at the left side of the gray box before typing the sub/superscript.

- Another way to make a subscript is to use square brackets - A[1] = A_1

Parentheses and Absolute Value Notation

You can add parentheses or an absolute value sign to a term in one of two ways:

- From the data input box, type the term(s), highlight it, then click the

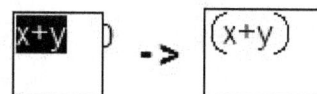

Parentheses (⁰) or **Absolute** Value |⁰| button.

- From the data entry box, click the **Parentheses** (⁰) or **Absolute** Value |⁰| button and enter the terms.

Built-In Functions

For including in any expression or constraint, *Gx* has the following common functions available: \

Trig

- sin()
- cos()
- tan()

- arcsin()
- arccos()
- arctan()

- sinh()
- cosh()
- tanh()

Math

- sqrt() - same as √⁰

- abs() - same as |⁰|

- signum(*x)* - finds the sign of a number:

$$= -1 \text{ if } x < 0$$
$$0 \text{ if } x = 0$$
$$1 \text{ if } x > 0$$

- log() or ln() - both mean the natural log

- diff(expression, variable)

- sum(expression, variable = start, end)

- ceil() - rounds up

- exp(x) - the exponential function; you must use *exp*, <u>not</u> *e, e* is just a variable name

- piecewise({expression1, domain1},{expression2, domain2}...) - the function is evaluated in the order written

- integrate(expression, variable)

- pi - same as π in the Symbols toolbox

- floor() - rounds down

Boolean

- AND
- OR
- NOT

- <
- <=
- >
- >=

Annotation Symbols

Using Annotation Symbols

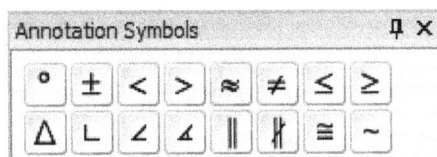

Annotation Symbols are available for use with any of the **Annotate** tools. These symbols are not used in *Math Illustrations'* algebra engine, but may be useful in creating worksheets and tests.

Use the symbols from the data entry window of the Annotate tools. Simply click the symbol you need as you type.

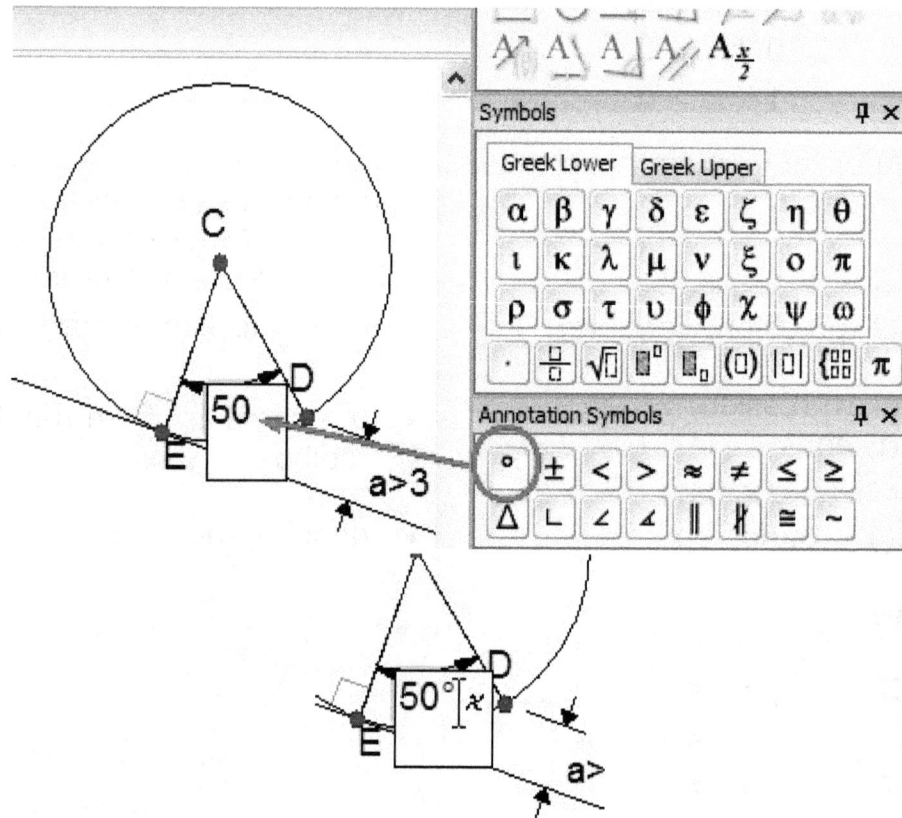

When you are finished with the annotation, press enter.

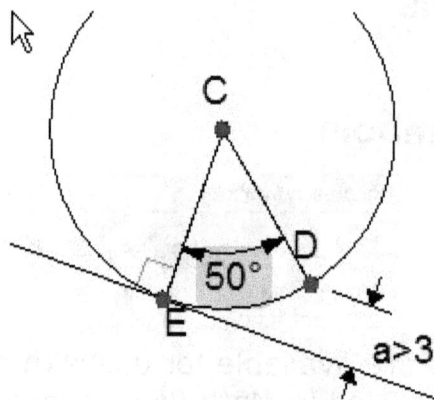

System Variables and Animation

Investigating Variables

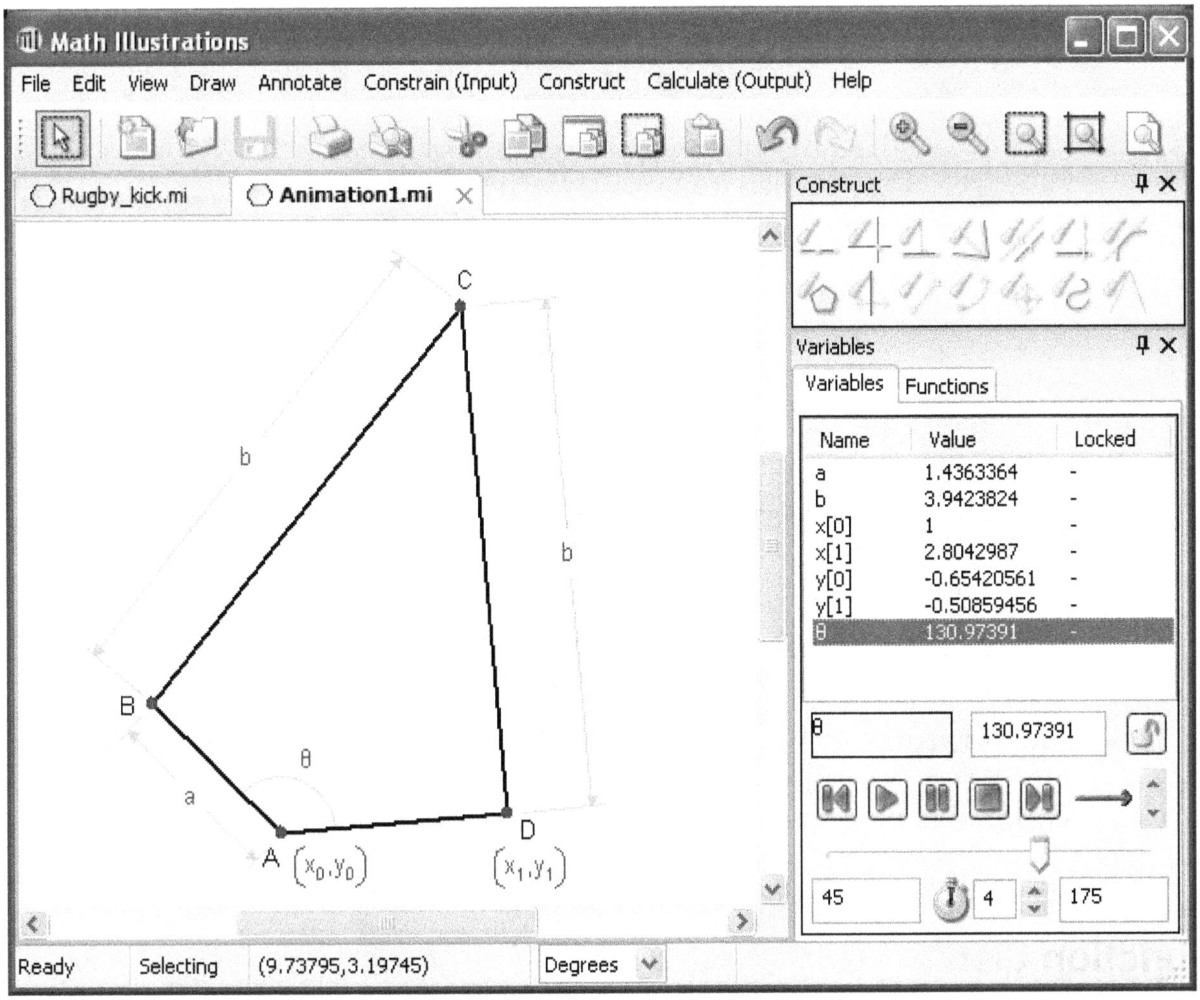

The **Variables** toolbox reports all the variables you have used in the diagram and lets you manipulate their values.

Variable List

Variables	Functions	
Name	Value	Locked
a	1.4363364	-
b	3.9423824	-
x[0]	1	-
x[1]	2.8042987	-
y[0]	-0.65420561	-
y[1]	-0.50859456	-
θ	130.97391	-

θ   130.97391

45   4   175

This list contains the names of all variables used in your diagram.

For every variable name, the system shows:

- **<u>the current value</u>** - these values can be ones that you have explicitly specified, or just taken from the way you sketched the geometry.

- **<u>lock status</u>** - if the variable is locked (+) its value will not change if you move the geometry or add additional constraints; the unlocked (-) variable is free to change as the geometry moves or changes.

Function List

When using the Function command to draw a function of the form Y = f(X) + g(X), the **Functions** tab in the **Variables** toolbox lists the functions *f* and *g* and their values. Use the edit line at the bottom of the box to modify the functions.

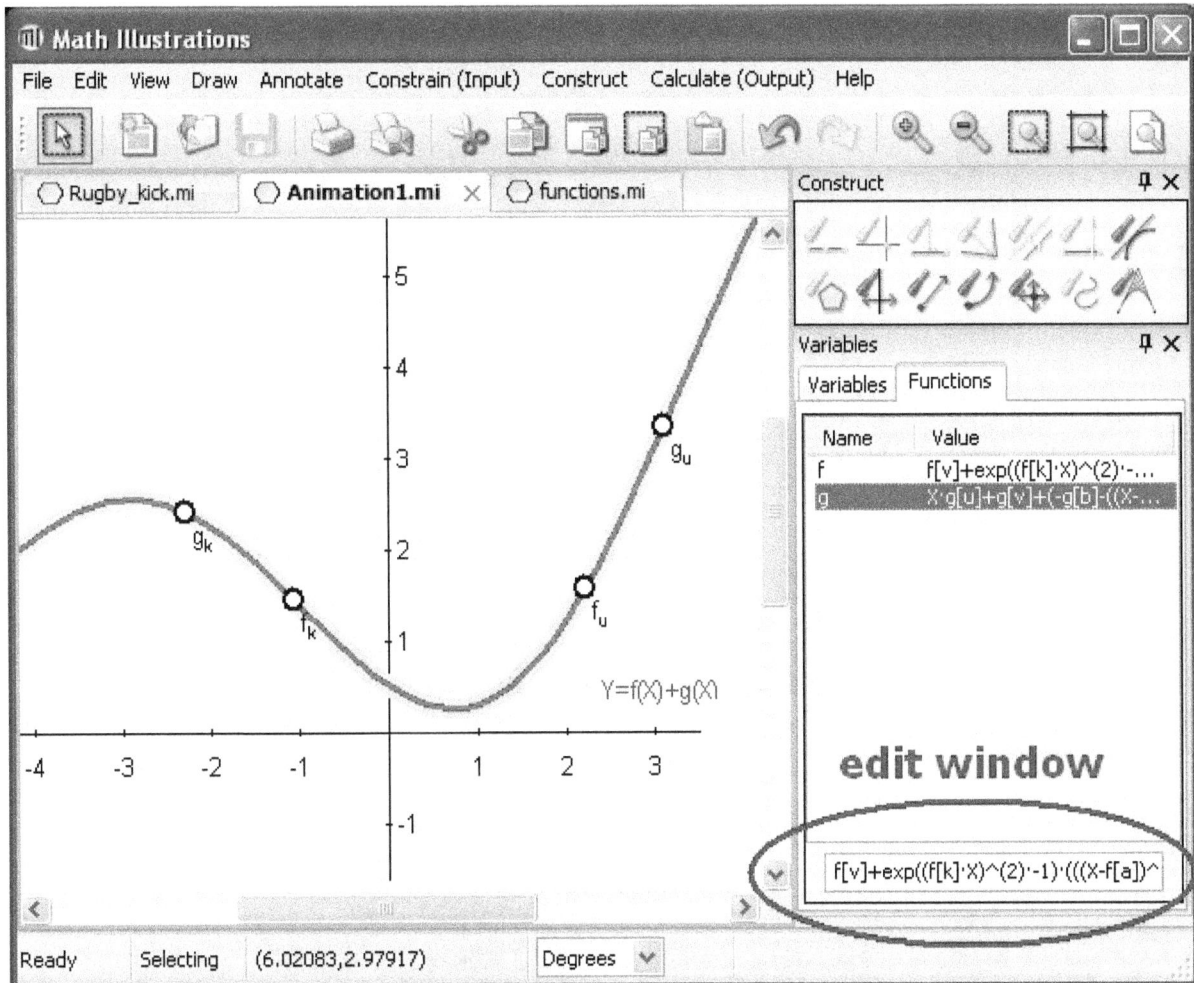

Using the Lock Tool

By default, when you drag points in a *Math Illustrations* model, it will adjust the numerical sample values used in the various parameters of the model to accommodate the drag, as best it can.

For example, in the model of a 4 bar linkage below, dragging point B will cause lengths a and b and angle θ to be adjusted appropriately.

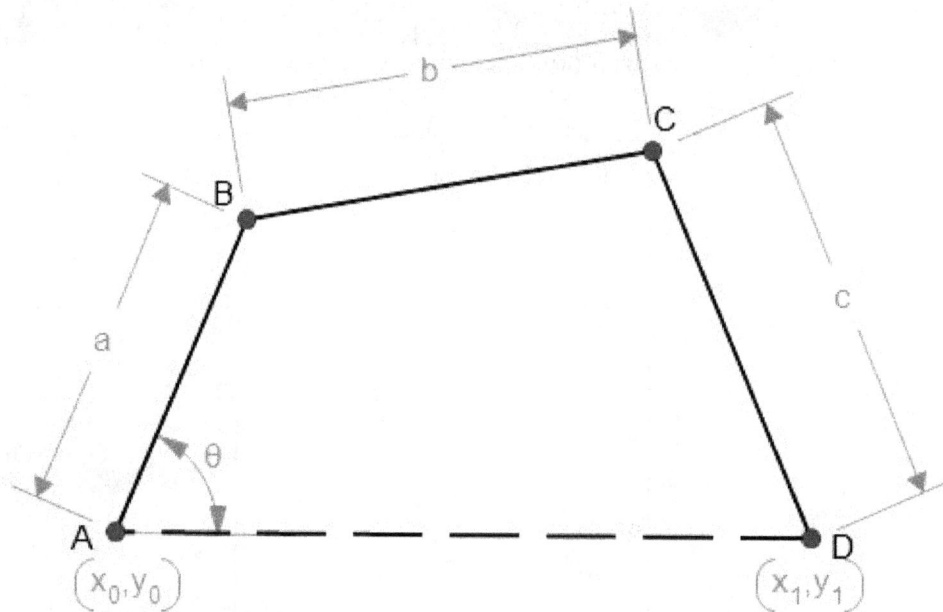

However, you may want the drag to act as if the members AB and BC were rigid, and only angle theta adjustable. To do this you can lock the parameters:

The value of a, for example, can still be set from the Variables panel, but it will not change when the model is dragged.

Changing and Locking the Variable Value

θ | 117.055 | 🔒

To make a change to the variable list, first click anyplace in the row of the variable you want to change. That row will be highlighted.

To change the value: highlight the value in the edit window and type the new value.

To change the lock status: just click the button -

🔓 to unlock a locked variable

🔒 to lock an open variable

Animation

Your geometry comes to life with the Animation tools. You simply need to select the parameter that drives the animation, give it a range, then Play.

In the diagram below we select θ for the crank of this linkage.

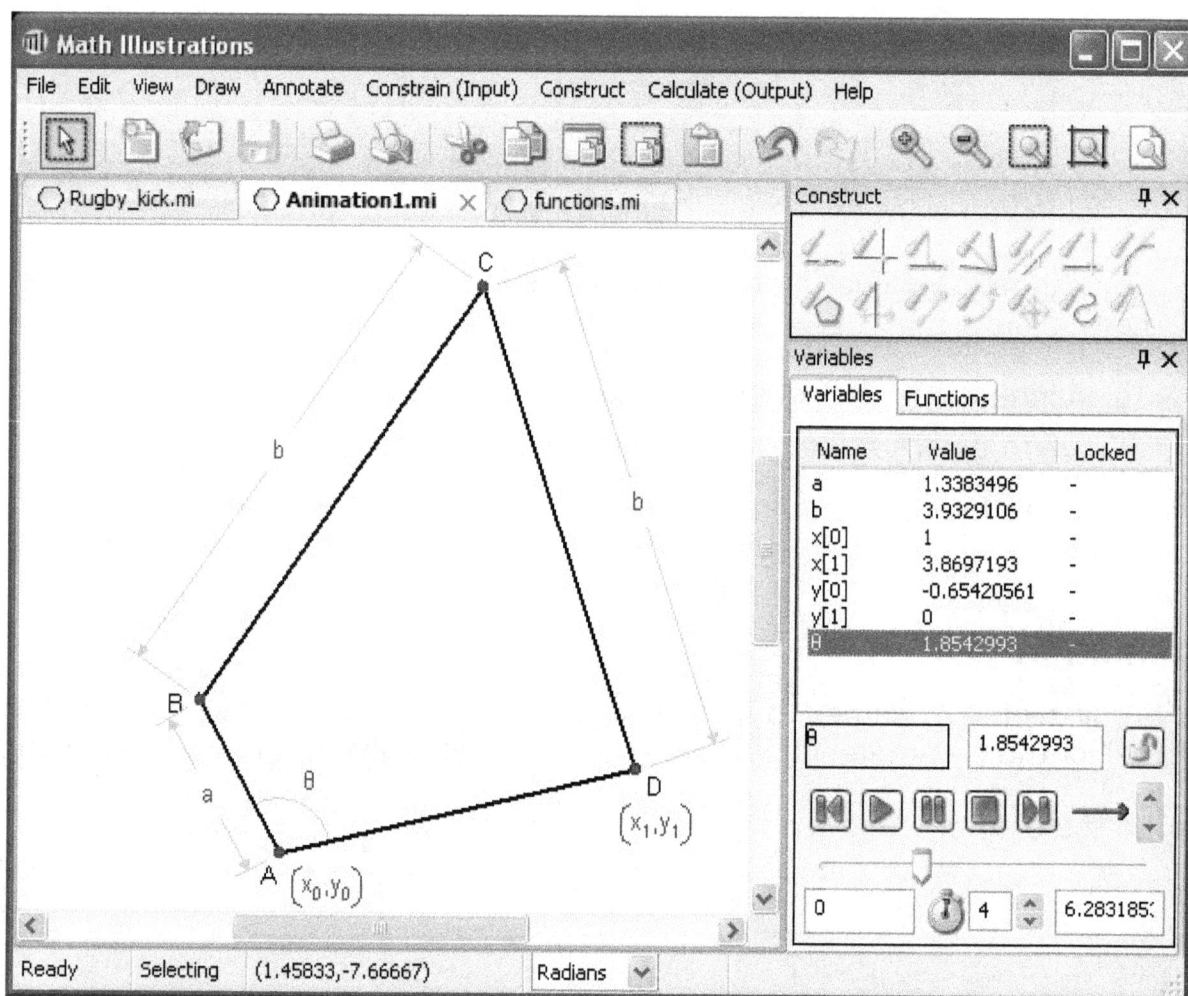

Click the headings below for details on the animation buttons and windows:

- Animation console - works like a video player.

- Animation modes - indicates how the range for the animation is stepped through.

- Animation values and duration - where you specify the speed and the range for the driving parameter.

Users of numeric interactive geometry systems may be familiar with the concept of animation based on points animated along line segments or curves. This type of animation can be conveniently modeled in **Math Illustrations** using the point proportional along a curve constraint along with parameter based animation.

Animation Console

The Animation console works like a standard video console with the **Play**, **Pause**, and **Stop** buttons as well as advance to the **Beginning** and **End** buttons.

Animation Modes

The animation modes can be changed with the up/down arrow buttons. The modes are:

Runs the animation one time through the specified range.

Runs the animation continuously from the beginning to the end of the range.

Runs the animation one time forward and then backward through the specified range.

Runs the animation continuously forward and then backward through the specified range.

Animation Values and Duration

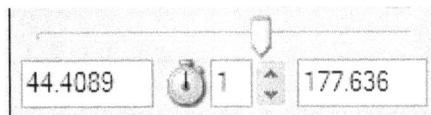

| 44.4089 | 1 | 177.636 |

These animation buttons help you adjust the range and speed of the animation.

- Click and drag the slider along the bar to manually animate the drawing.

- In the two data entry windows at the bottom right and left of the toolbox, specify the range of the animation.

- The **Duration** box in the center lets you specify how long the animation takes to play one time through. Values are between 1 and 60 seconds.

Animation and the Locus Tool

Both the construction of the locus and envelope curves, and the animation of the diagram in *Math Illustrations* can be defined in terms of any variable. For example in the model below, we can create a locus over values of the variable t (other variables will be kept constant).

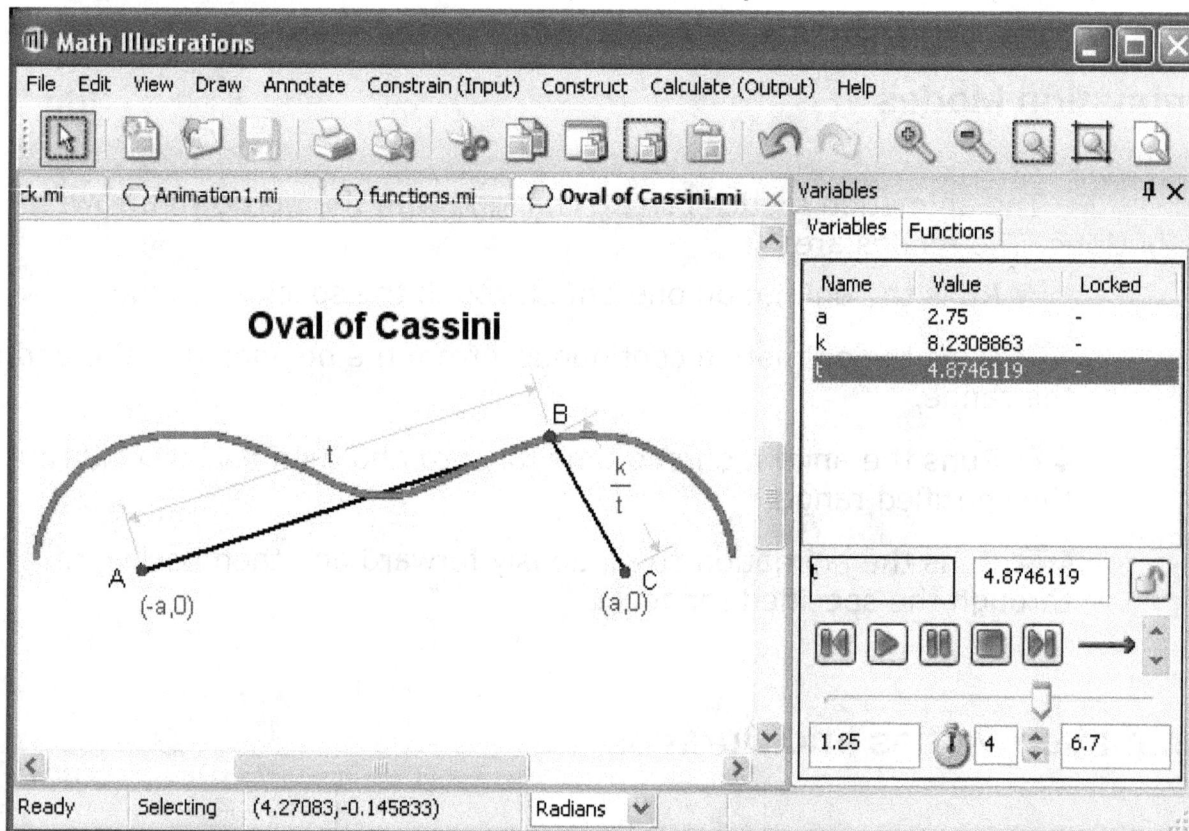

Menus and Icons

Menus

File Menu

The **File** menu contains the standard Windows file handling operations with options for copying and exporting to other programs. Several of the options are also available from the icon bar.

Menu Option	Function
New	Creates a new project.
New Graph	Graph mode allows scaling of axes.
Open. . .	Brings up the **Select a File** dialog box so you can open a project.
Close	Closes the current file or, if multiple files are open, the file on top.
Save	Saves the file. If you have not yet saved the current work to a file, the **Save File As** dialog box lets you specify where to save the project file.
Save As . . .	Brings up the **Save File As** dialog box so you can specify where to save the project file or change the file name.
Open Workbook	Brings up the **Open Workbook** dialog box. If any other files are open, they will be closed when you select the workbook.
Save Workbook	Saves all tabbed pages as a single workspace (.miw), so you can open them all at once.
Save Workbook as	Saves all tabbed pages as a single workbook (.miw) and lets you specify the path and file name of the workbook.
Close Workbook	Closes the current workbook. If you have made changes to pages, you will be asked if you want to save them individually (.mi files).
Import Figure from Figure Gallery	A shortcut containing common geometry figures that you may need for creating worksheets or other documents.
Export	Export the file as Windows Metafile (.emf) (**Windows version only**), an Encapsulated PostScript (.eps), a scalable vector graphics (.svg), an image, or an animated gif.

Page Setup. . .	Displays the page setup dialog box for choosing a printer and print options.
Print Preview	Displays the printout by pages.
Print. . .	Displays the standard system Print dialog.
Recent files	Click to display a list of the most recently used files. Selecting one opens it.
Exit	Exits the program, after prompting for save.

Graphing Mode

When you need to draw a graph requiring independent scaling of the axes, select **File / New Graph**.

- Draw your graph.
- Click an axis - the axis will be highlighted and a circular handle appears.

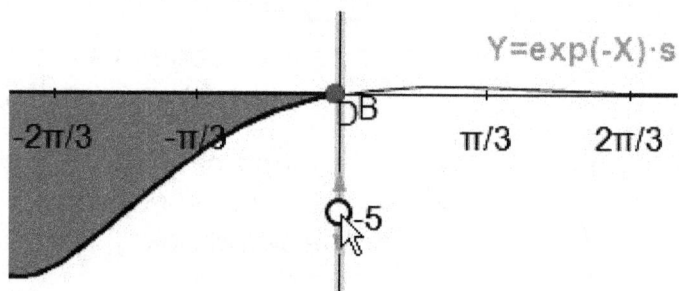

- Slide the handle up and down the axis with your mouse.

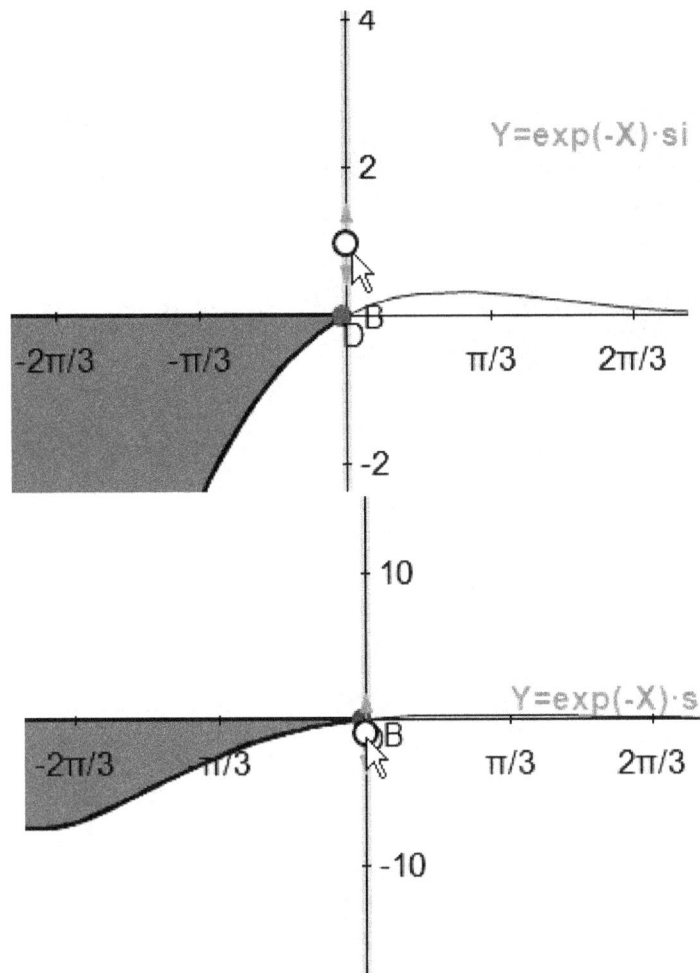

Note: Many of the geometry drawing, constraint and construction tools are unavailable (grayed out) in the Graphing mode for obvious reasons related to the independent scaling of the axes. And you cannot copy and paste between geometry and graph documents.

Importing Files from the Figure Gallery

Are there figures which you need to use frequently? The Figure Gallery makes your tasks easier. Browse through the folders to see the many objects, graphs and transformation examples which you can use and tailor to your needs without starting from scratch.

Figure Gallery files are copied to your computer when you install *Math Illustrations*.

Here are the steps to import a figure:

1. Select **File / Import Figure from Figure Gallery**

2. Double click a folder or sub-folder, or click and press Select in the lower right corner to view the problems in a category. You can also use the search window to find files.

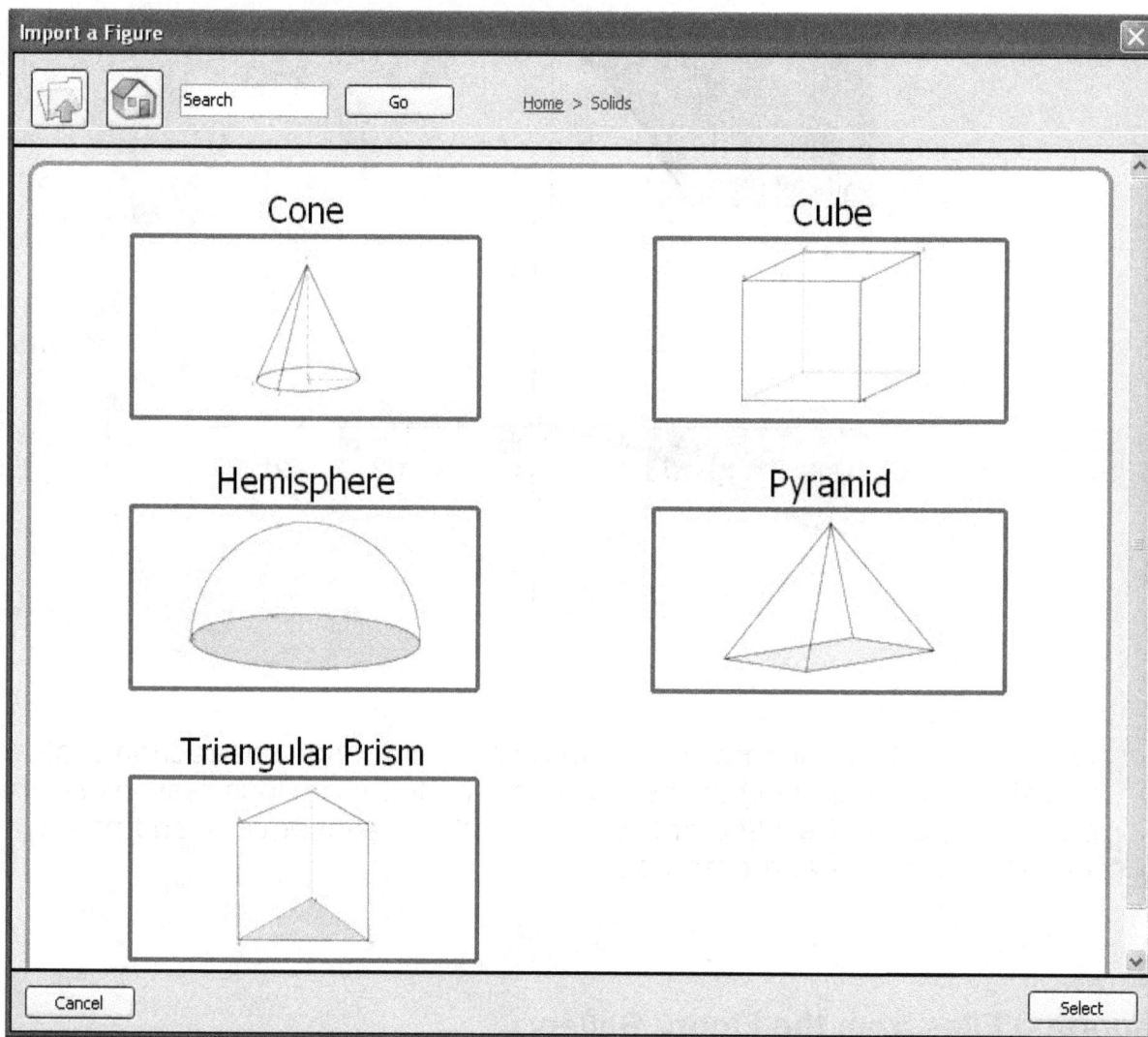

3. Double click a problem (or click and press Select) to see an enlarged view.

4. Click the Import button (it replaces the Select button in the lower right corner) to bring the drawing into your drawing window.

Use the Toggle Hidden function from the general context menu to modify constraints in the drawing, or try dragging the geometry to suit your needs. Don't forget to Save.

Navigating the Figure Gallery

The Figure Gallery is arranged in the usual tree structure containing folders and sub-folders.

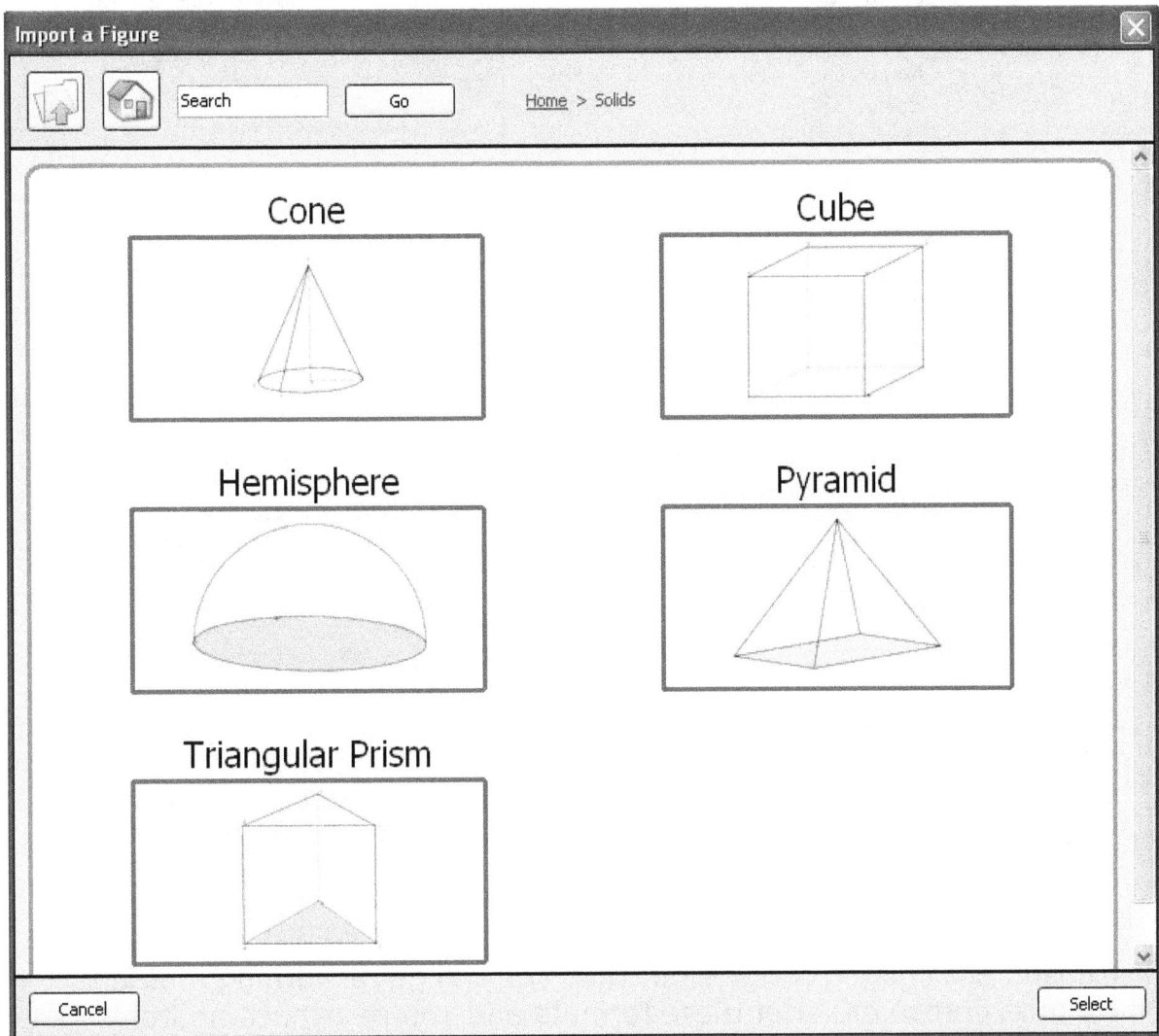

Back - takes you up one level.

Home - takes you to the top level.

Use the Search window to find a specific file or types of files. Enter the search word(s) and click Go.

Text at the top center of the dialog tells you which level is displayed. Click Home to return to the top level.

Exporting a Drawing

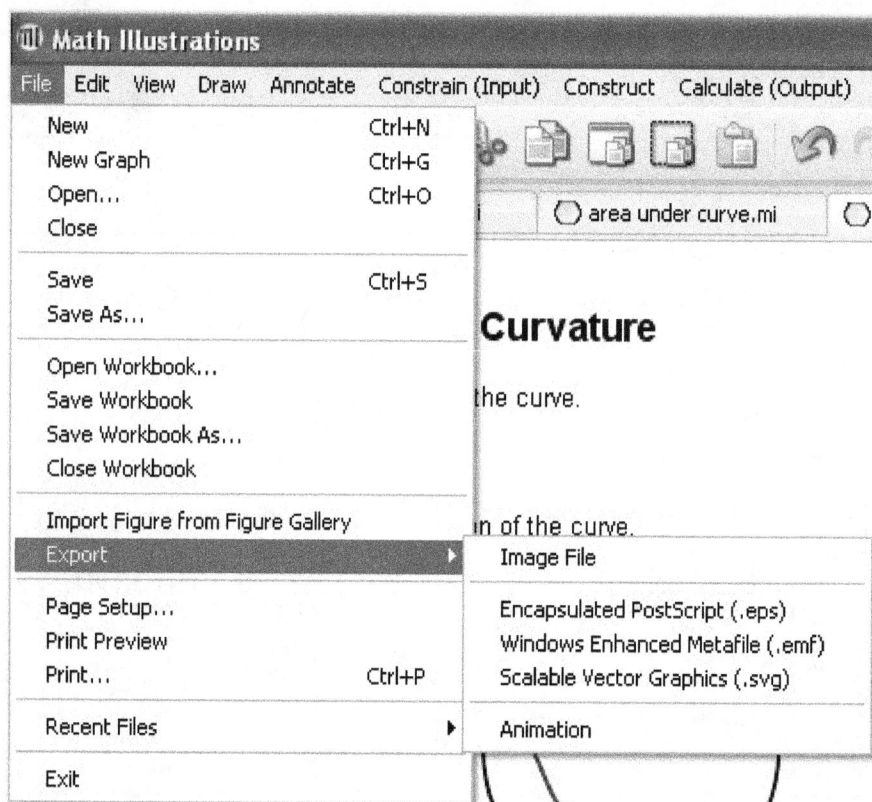

To export your drawing to another program, choose **File / Export**. You can export the drawing as a *Windows Enhanced Metafile, Encapsulated Postscript, Scalable Vector Graphics,* in one of several standard image formats, or as an animated *gif*.

Please note that neither EMF, nor EPS support semi-transparency or transparent images of any kind, thus you will get a warning message if you try to export to either of these formats and you have such an item in the document.

Here are the steps for the export:

1. For all file types enter the Filename or click the folder icon to select the appropriate folder and file.

2. Image and Animation files have an extra step at this point as detailed below.

3. Select the region of the drawing with the displayed cursor   (click

opposite corners of the region).

Image Files

When exporting image files, click the down arrow in the <u>Save as type</u> line to select your desired image format.

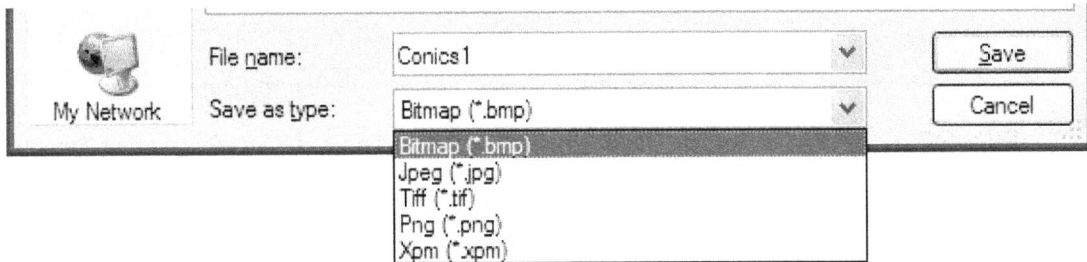

Next, set the resolution in the <u>File DPI</u> box. Click the down arrow and select the appropriate setting or enter a number in the window.

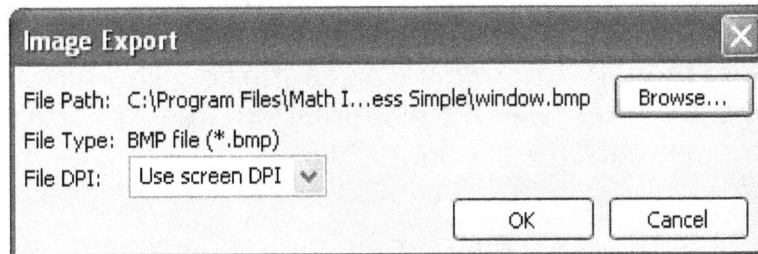

Animation Files

Select **File / Export / Animation File** to produce an animated *gif*. This format is supported by many applications and will enable you to embed animations in, for example, *PowerPoint* slides and *Wikipedia* pages.

Select the directory and file name of your *gif*, and you will be presented with a dialog to choose the parameter on which the animation is based, along with various technical aspects of the animation.

Animation Export

File Path:	C:\GeometryExpressions\reflection.gif Browse...
File Type:	Animated GIF
File DPI:	Use screen DPI
Frames Per Second:	10
Number of iterations:	0 Enter "0" for unlimited iterations.
Variable:	t

OK   Cancel

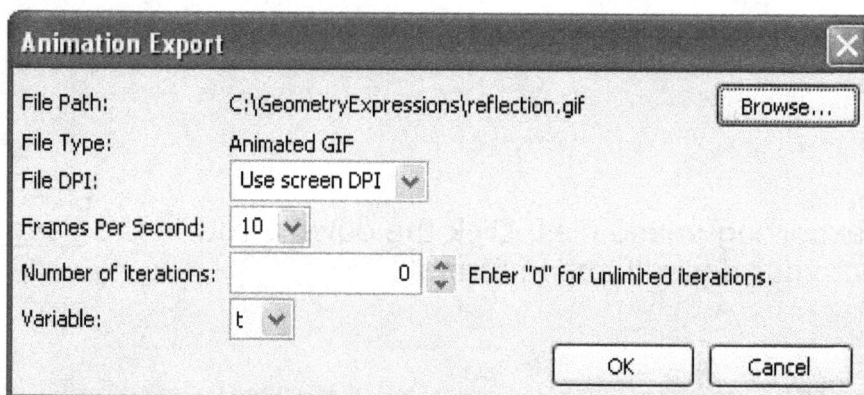

<u>File DPI</u> - specify the resolution of the output. The higher number you use, the slower will be the process of creating and loading the animation.

<u>Frames Per Second</u> - if you multiply this number by the animation duration specified in the **Variables** toolbox, you will get the number of frames captured. For example if you are set at 10 frames per second, and the **Variables** toolbox specifies the duration of the animation to be 4 seconds, then 40 frames will be captured. The more frames you capture, the slower will be the animation creation process, and the longer the animation will take to load.

<u>Number of iterations</u> - when an animation is played (*e.g.* when a *PowerPoint* slide containing the animation is displayed), enter a number to play the animation a specific number of times, or enter 0 to play it continuously.

<u>Variable</u> - choose the variable that controls the animation. (All the variables in the **Variables** toolbox should be available). The limits of the variable defining the range of the animation should be set in the **Variables** toolbox.

Select the region of the drawing with the displayed cursor   (click opposite corners of the region).

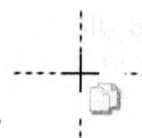

Edit Menu

The **Edit** menu contains the standard Windows editing operations as well as ways of dealing with constraint conflicts and all of the program settings. Several of the options are also available from the icon bar.

Menu Option	Function	When Available
Undo	Reverses actions starting with the last one.	After any action has been taken.
Redo	Reinstates actions starting with the last one that was undone.	After using **Undo**
Select	When checked, the select mode is active.	Select mode is always active except when using a Drawing tool or moving or panning the drawing.
Select All	Selects everything in the drawing window.	Always
Select All Type	Presents a submenu of object types to select.	Always -- most useful when the object type is in the window.
Clear Selection	Unselects any objects that are selected.	Always
Cut	Deletes an object, but saves it so it can be pasted somewhere else.	An object is selected
Copy	Does not delete the object, but saves is so it can be pasted somewhere else.	An object is selected
Copy Drawing	Copies everything in the drawing window as an Enhanced Metafile (.emf)	Always
Copy Region	Copies a section of the drawing window inside a rectangle made by dragging the cursor.	Always

📋 Paste	Puts whatever has been cut or copied into the current document	Object(s) cut or copied
Delete	Deletes whatever is selected, without saving it.	One or more object(s) selected
Arrange	For placing selected objects on top of (bring forward) or underneath (send backward) other objects for easier selection and viewing.	One or more objects selected
Convert to Measurement	Removes the selected constraint and replaces it with the corresponding measurement (output)	An input constraint is selected
Details. . .	Displays the Edit Text dialog to edit a block of text	A block of text is selected
Parameters. . .	Lets you edit the parameters of functions, loci and traces.	A function, locus or trace is selected
Properties. . .	Lets you edit the display properties of the selected object(s)	One or more objects of the same type are selected
Preferences. . .	Sets the default appearance and properties for the project's drawing, text and mathematics	Always

NOTE: For the Mac version, **Preferences...** is listed under **Math Illustrations** menu.

View Menu

The table below lists the complete summary of **View** functions.

Menu Option	Function	When Available
Hide	Hides a selection	One or more element(s) are selected.
Show all	Displays any entities that were hidden	One or more element(s) are hidden.
Toggle Hidden	Lets you toggle hidden / visible for any object in the drawing	Always (if the drawing window isn't empty)
Zoom In	Makes the drawing details larger without affecting the size on the printed page. (The text gets larger on the screen.)	Always available - (most useful when there is something in the drawing window).
Zoom Out	Makes the drawing details smaller without affecting the size on the printed page. (The text gets smaller on the screen.)	Always available - (most useful when there is something in the drawing window).
Zoom To Selection	Lets you make a selection and adjusts it to fit the drawing window. (The text gets larger on the screen.)	Always available - (most useful when there is something in the drawing window).
Zoom To Fit	The entire diagram is displayed in the drawing window. (The text size changes with the geometry.)	Always available - (most useful when there is something in the drawing window).
Zoom To Page	The whole page is displayed in the drawing window. (The text size changes with the geometry.)	Always available - (most useful when there is something in the drawing window).
Pan View	Allows you to move the contents of the drawing window without changing its position on the page.	Always available - either **Pan View** or **Move Geometry** will be in effect (checked).

Scale Geometry Up	Enlarges only the geometry. (The text size on the screen doesn't change.)	Always
Scale Geometry Down	Shrinks only the geometry. (The text size on the screen doesn't change.)	Always
Scale Geometry To Selection	Lets you select a portion of the geometry and adjusts it to fit the drawing window (The text size on the screen doesn't change.)	Always
Scale Geometry To Fit	Adjusts all geometry to fit in the drawing window. (The text size on the screen doesn't change.)	Always
Scale Geometry To Page	Adjusts all geometry to fit inside the specified page boundaries. (The text size doesn't change relative to the page.)	Always
Move Geometry	When checked, click and drag to move the drawing contents with respect to the page boundaries.	Always. (Make sure **Page Boundaries** is checked to see the results.)
Axes	When checked, the axes are displayed. They have the properties of Infinite Lines.	Always
Grid	When checked, the grid is displayed.	Always
Page Boundaries	When checked, the page boundaries are displayed.	Always
Tool Panels	Lists all the toolboxes. When checked, the toolboxes are displayed on the screen.	Always

Tool Panel Configurations	Gives you options for arranging the toolboxes to your preference.	
Language	The current version of *Math Illustrations* can be displayed in **English**, **French**, **German**, **Spanish**, **Polish**, or **Russian**. Choose one and restart the program.	Always

Checked menu options are toggles:

- Checked indicates the option / mode is active or displayed.

- Unchecked indicates the option / mode is inactive or hidden.

- Except **Pan View** and **Move Geometry** where one or the other is checked.

Click the selection to change its state.

Some menu items have icon shortcuts found on the icon bar at the top of the screen.

Zooming and Scaling

The **View** menu has **Zoom** operations pertaining to the screen view, and **Scale** operations pertaining to the page view.

- Zooming makes the drawing details smaller without affecting the size on the printed page. The text (constraints, output and annotation) changes size with the rest of the drawing.

- Scaling adjusts the size of the geometry relative to the page, but the text doesn't change size in the drawing window. Check **View / Page Boundaries** to see this work.

The **Scale** functions used from the icon bar at the top of the screen can be changed to **Zoom** functions by holding the ctrl key while clicking the icon. This is handy if you need to change the size of the text on the screen:

- A **Scale down** followed by a **Zoom in** [ctrl] has the effect of enlarging the text.

- A **Zoom out** [ctrl] followed by a **Scale up** has the effect of shrinking the text on the screen.

Toolbox Menus

The menus with the same name as the toolboxes at the side of the screen just give another way of accessing the same functions.

Menu Option	Selection		
Draw	Point	Line Segment	Infinite Line
	Vector	Polygon	Circle
	Ellipse	Parabola	Hyperbola
	Arc	N-gon	Curve Approximation
	Text	Picture	Expression
	Function		
Annotate	Angle	Distance / Length	Coefficients
	Coordinate	Radius	Expression
	Direction	Slope	
Constrain	Distance / Length	Radius	Perpendicular
	Angle	Direction	Slope
	Coordinate	Coefficients	Tangent
	Incident	Congruent	Parallel
	Equation	Proportional	

Construct	Midpoint	Intersection	Perpendicular Bisector
	Angle Bisector	Parallel	Perpendicular
	Tangent to Curve	Polygon	Reflection
	Translation	Rotation	Dilation
	Locus	Trace	Area Under Arc
Calculate	Distance / Length	Radius	Angle
	Direction	Slope	Coordinates
	Area	Perimeter	Coefficients
	Parametric Equation	Implicit Equation	

Help Menu

The **Help** menu lets you access this help system, check for updates, change the program's language, and gives you information about the program's license and version. The menu selections are always available.

Menu Option	Function
Dynamic Help	Invokes the Help system
Contents. . .	Look in the Table of Contents; add new or refer to saved bookmarks.
Index. . .	Look in the **Help** index. There is also a facility to Search index headings.
Search. . .	Search the **Help** topics for keywords.
License. . .	Displays information about your license.
Check for Updates. . .	Prompts you to save your work, checks for new versions of Math Illustrations, then restarts the program.

About. . .	Contains the current version of the program, the copyright notice, and the link to Math Illustrations' website.

NOTE: For the Mac version, **About...** is listed under **Math Illustrations** menu.

Context Menus

Context Menus pop up when you right-click with the cursor positioned anywhere in the drawing window.

- The general context menu - appears when you right click and nothing is selected.

- The selection context menu - appears when one or more elements in the drawing window are selected. Some menu entries my be inactive, depending on which elements are selected.

The General Context Menu

Right-click anywhere in the Drawing Window to display a context menu. If nothing in the window is selected, the menu choices are the following:

Menu Option	Function	When Available
Close	Closes the current file	Always
Save	Updates a file that already exists	The file has been saved
Save As. . .	Saves a file for the first time and prompt for the filename and path	Always
Select All	Selects everything in the drawing window	Always
Select All Type	Presents a submenu of object types to select	Always -- most useful when the object type is in the window

Copy Drawing	Copies everything in the drawing window as an Enhanced Metafile (.emf)	Always
Copy Region	Copies a section of the drawing window inside a rectangle made by dragging the cursor.	Always
Paste	Puts whatever has been cut or copied into the current document	Object(s) cut or copied
Show All	Displays any entities that were hidden	One or more element(s) are hidden
Toggle Hidden	Lets you toggle hidden / visible for any object in the drawing	Always (if the drawing window isn't empty)

Toggling - Hide / Show Elements

From the general context menu select Toggle Hidden. The magic wand cursor appears, and any hidden objects appear faintly in the drawing window.

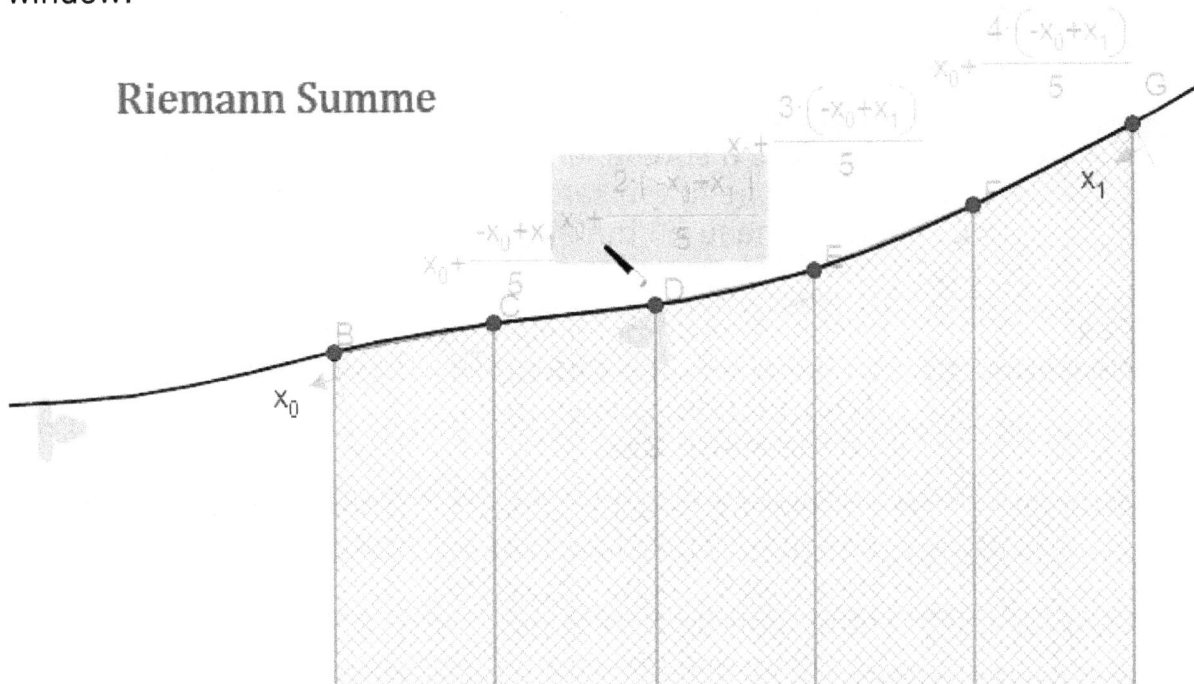

Click any items that are faint to display them. Click any displayed items to hide them. When you are finished toggling, click the select arrow ⬚.

Selection Context Menu

Menu Option	Function	When Available
Cut	Deletes an object, but saves it so it can be pasted somewhere else.	One or more objects selected
Copy	Does not delete the object, but saves is so it can be pasted somewhere else.	One or more objects selected
Paste	Puts whatever has been cut or copied into the current document	Object(s) cut or copied
Delete	Deletes whatever is selected, without saving it.	One or more objects selected
Hide	Makes the selected objects invisible	Makes the selected objects invisible
Constrain (Input)	Displays a submenu identical to the one in the same drop-down menu on the **Menu Bar**	A geometry object is selected
Construct	Displays a submenu identical to the one in the same drop-down menu on the **Menu Bar**	A geometry object is selected
Real Measurement	Displays a submenu identical to the one in the **Calculate (Output)** drop-down menu on the **Menu Bar**	A geometry object is selected
Arrange	For placing selected objects on top of (bring forward) or underneath (send backward) other objects for easier selection and viewing.	One or more objects selected

Visibility Condition	Lets you enter an equation specifying when the selected object(s) are visible.	One or more objects selected
Point Properties	Lets you change the selected point's color and size and enable the Point Trace	One or more points selected
Arrow Head	Lets you turn off or change the arrow head style of the selected segment(s) or arc(s)	One or more line segments or arc(s) selected
Arrow Size	Lets you choose the size of the arrowheads (when used)	One or more line segments or arcs with arrowheads selected
Line Properties	Lets you change the selected line's color, style, and thickness	One or more objects created with lines or segments
Fill Properties	Lets you change the color, style, and transparency level of the selected object(s)	Polygon, N-gon, filled circle, filled ellipse, picture or trace object (s) selected
Pin	Lets you pin or unpin the selected object(s)	Text, picture, or expression selected
Text Properties	Lets you change the selected text's color, size, and style	Text, label, expression, constraint, or measurement selected
Show Symbol	A toggle to display or hide the selected object(s)	A constraint, annotation, or measurement lines selected
Output Properties	Lets you turn on or off the selected output's name	One or more outputs selected
Show Arrowheads	Lets you turn on or off the selected angle symbol's arrowheads	One or more angle symbols selected
Congruence Style	Lets you change the angle style from arcs to tics and vice versa	One or more angle or congruent angle annotations selected

Tic / Arc Count	Lets you change the number of tic/arc counts of selected annotations	One or more congruent, congruent angle, or parallel annotations selected
Axis Arrow	Lets you put arrowheads and / or axis name labels on the selected axes	One or both axes selected
Axis Properties	Lets you turn the labels on the axes on/off or change the number of subdivisions or units to display	One or both axes selected
All Properties. . .	Lets you edit the display properties of the selected object(s)	One or several similar objects are selected
[Convert to Calculation (Output)]	Deletes the selected constraint and calculates the equivalent output	A constraint is selected

Visibility Condition

You can set any mathematical condition for one or more object's visibility. Use this with the Animation tools for some great effects. Here are the steps:

1. Select the object(s) that you want to change visibility.

2. Right-click and select **Visibility Condition** from the **Selection Context** menu.

3. Enter the expression for a defined variable for which you would like your object(s) to be visible.

Here's an example. Point C is t proportional along the parabola. In the first figure the picture is hidden. When the picture was visible, we set it's **Visibility Condition** to: $|t| >= 0$ AND $|t| < .3$. With the Animation tools we set t from -1.5 to 1.5. As point C approaches the top of the arc, BOOM!

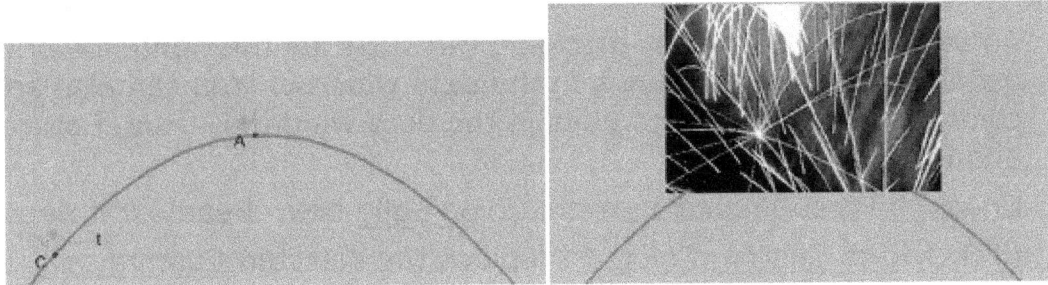

Editing the Color List

Use the **Context** menu to chang colors of any elements in your drawing. When you select **Point / Line / Fill / Text Color** the 16 color pallet appears. If you don't like the color choices you can edit the list in 2 ways.

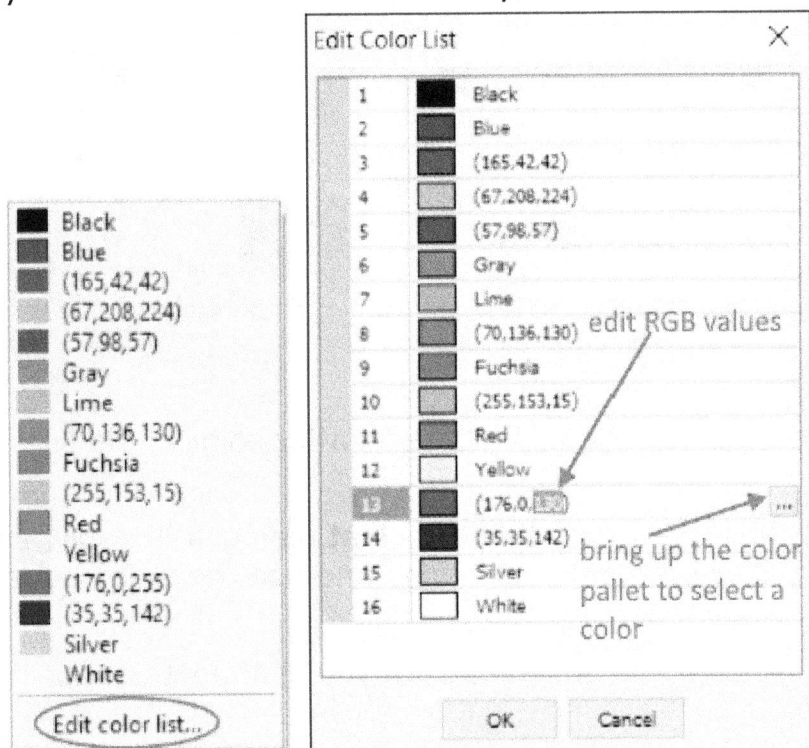

First click the <u>Edit color list</u>, then:

- change the RGB values in the list or
- click the ⸽ to bring up the color pallet and select a replacement color

Here are the steps for creating a color in the color pallet and replacing it in the list:

1. Click the color to be replaced from the <u>Custom colors</u>.

2. Move the cursor (click-and-drag) in the pallet to select your color
3. Adjust the Luminance by dragging the slider on the right side of the pallet or changing the value in the data window. You can also adjust the color by changing other values in the data windows - Hue, Saturation and RGB.
4. Check the resulting color in the Color/Solid box. Readjust if necessary.
5. Click Add to Custom Colors to replace the selected Custom color.

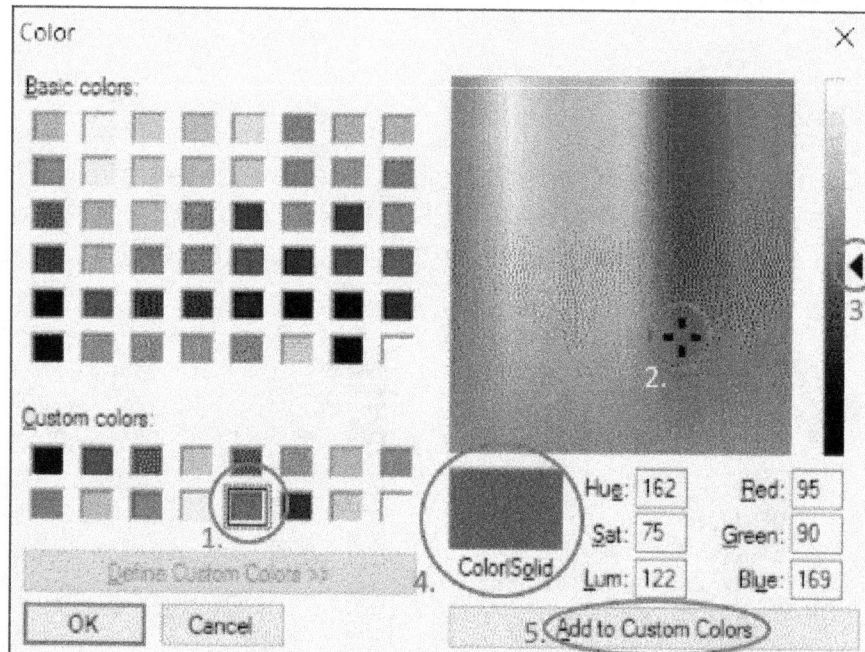

You can return to step 1. to change another color in the list.

Click OK to return to the **Edit Color List** dialog with your new replacement color(s), or Cancel change your mind and use the Color List as it was.

The new Color List will be saved until you change it again.

Tracing Points

You can select one or more points to see a trace of their path when dragging or moving the diagram. Here are the steps:

- select the point(s) to trace
- right-click to get the **Context** menu
- Select **Point Properties / Point Trace / Yes**

Click-and-drag or animate the drawing to see the trace. When you click again (or animate again) the last trace disappears and a new one is created. The difference between constructing a Locus and drawing a **Point Trace** is the **Locus** creates a mathematical expression, a **Point Trace** does not.

A nice example of the **Point Trace** is the pantograph.

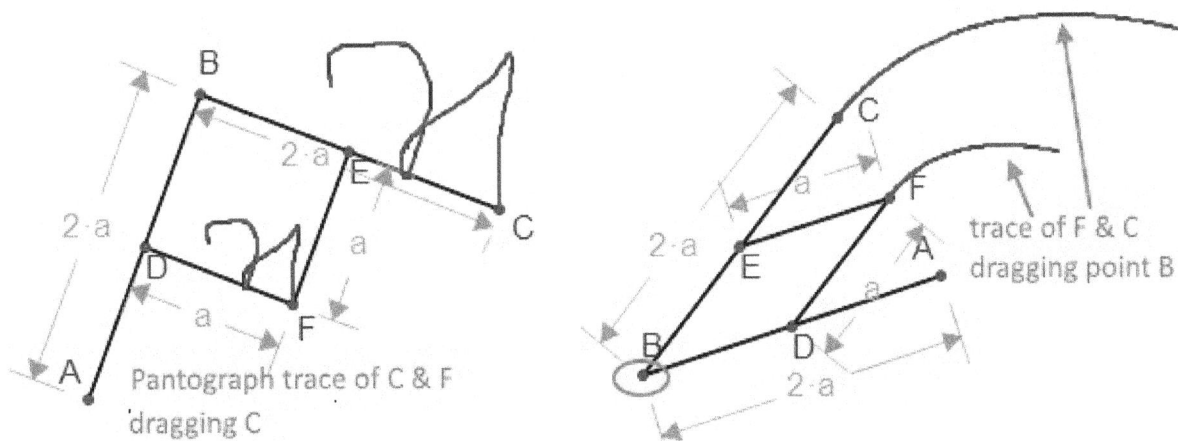

Pantograph trace of C & F dragging C

trace of F & C dragging point B

The trace color can be changed from the Context menu under **Line Properties / Line Color**.

Axes Display Properties

The Axes Display can be set in the default settings, **Edit / Preferences** under the **Grid, Axis, Page** tab, or you can change them for an individual drawing from the Selection Context menu.

To invoke the **Selection Context** menu:
1. Select one or both axes
2. Right click the mouse

Submenu choices include:
- **Axis Arrow** - lets you put arrowheads and / or axis name labels on the selected axes.

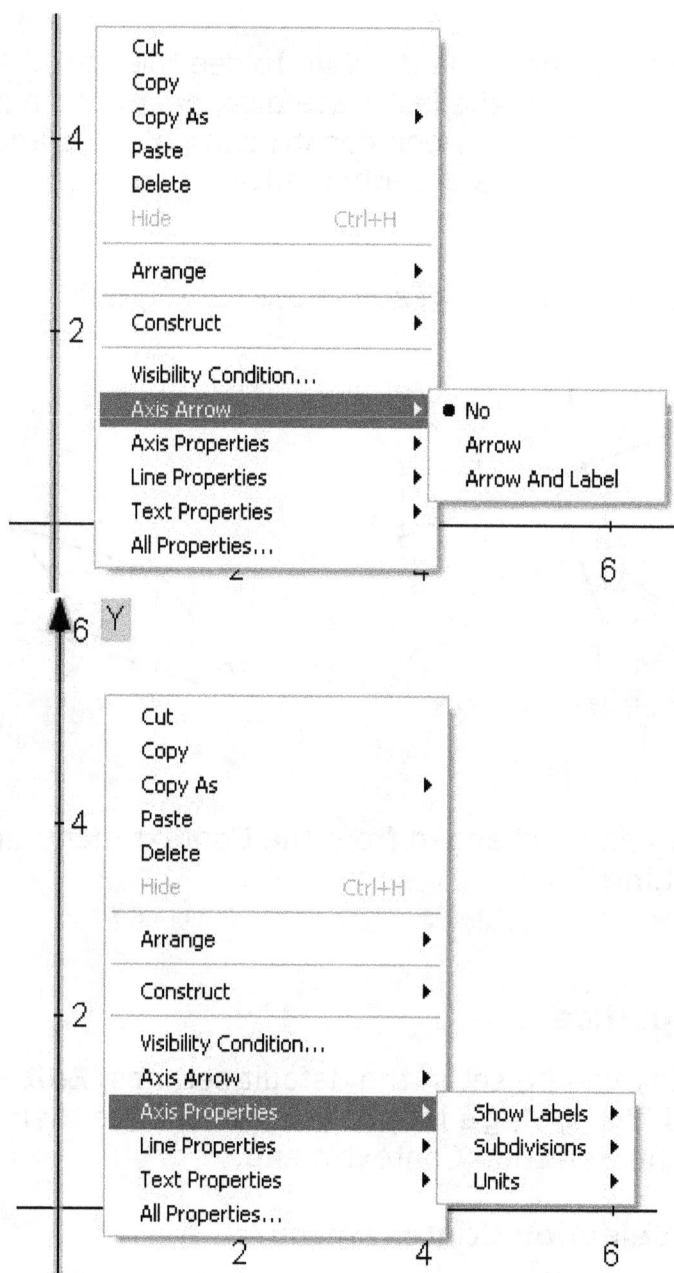

Display Properties

Axes	
Line Color	■ Black
Line Style	Solid
Line Thickness	1
⊞ Font	10; Swiss; Arial; Normal; Normal; Not Underlined; Black
Show Labels	☑
Units	Decimal ⌄
Subdivisions	Decimal
Visibility Condition	Degrees
Axis Arrow	Radians
	Radians/3

OK   Cancel

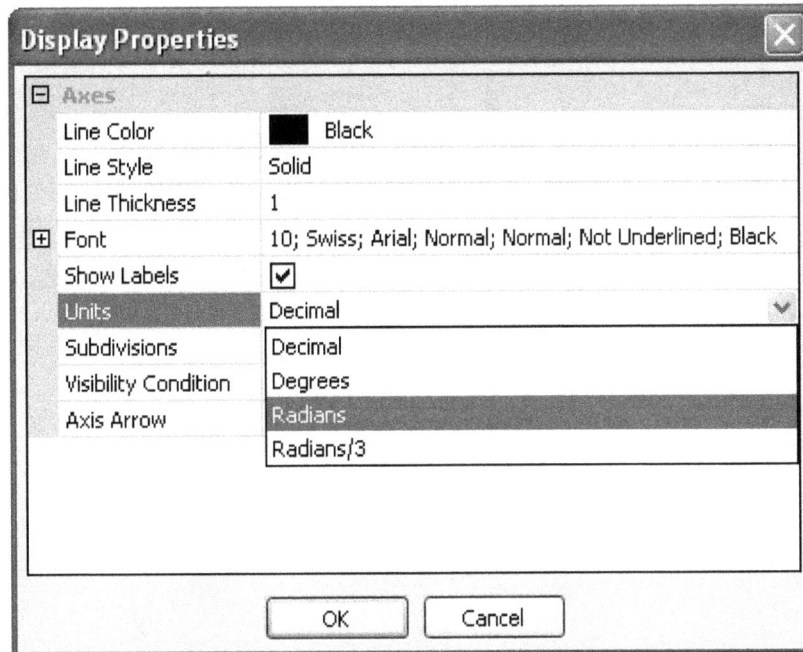

You can configure the X and Y axes in many ways. There are the usual attributes for lines and text.

- **Show Labels -** these are the unit labels which can be selected as shown in the dialog above. You may want to hide these if they obscure part of your drawing.
  - To turn these off select one or both axes and click the checkbox (unchecked).
  - To turn labels on again, make your selection and click the box to display the checkmark (it's a toggle).

- **Units** - can be set to the usual decimal, but with trig functions you may find the Degrees or Radians units more appropriate.

No matter what your units, for trig functions, don't forget to set your Angle Mode to **Radians** on the status bar.

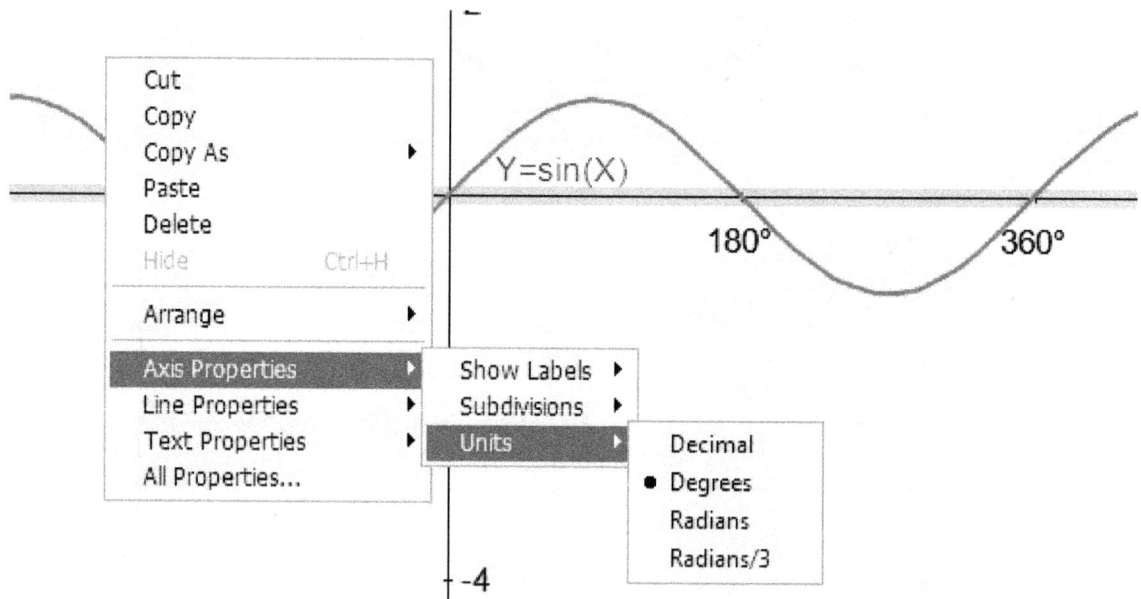

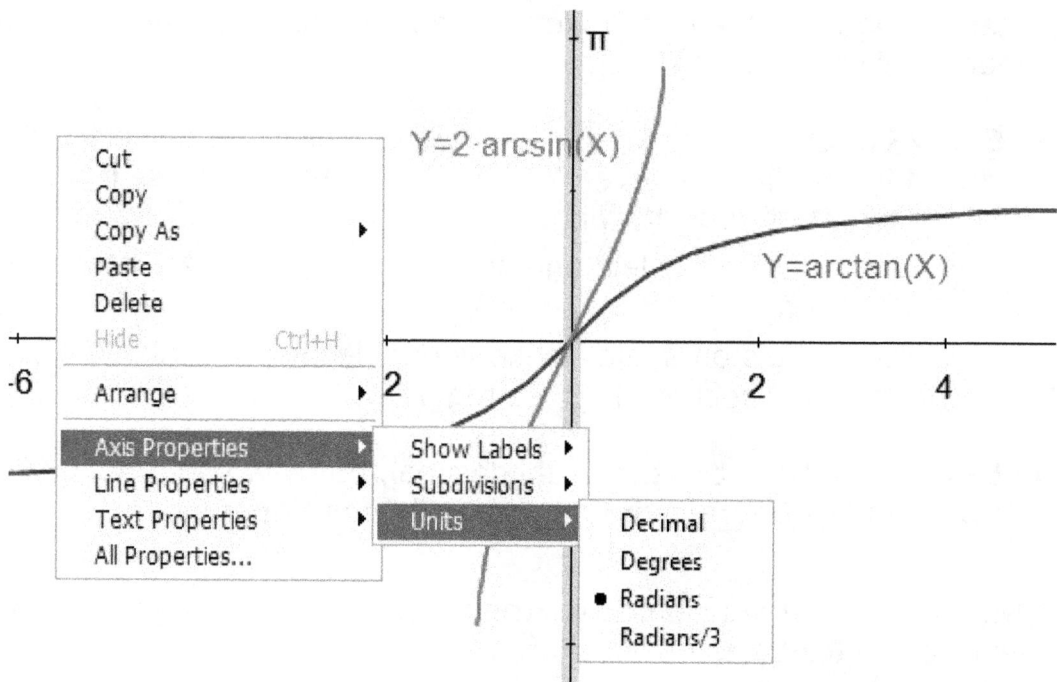

Icons

Tool Bar

The icons across the top of the screen make some of the routine tasks in the **File, Edit,** and **View** menus, and **Help** easily accessible.

- See **The Selection Arrow** for more information

- See **File Menu** for more information

- See **Edit Menu** for more information

- See **View Menu** for more information

Invokes this Help facility

More About Math Illustrations

What's New in Math Illustrations?

New in Math Illustrations v3.1

Features our users have requested -- we aim to please!

- More Arrowheads: we gave you arrowheads on line segments, now you can put them on one or both ends of arcs.

- Better Arrowheads: and now you can select the size of your arc and segment arrowheads, all provided for in the Context menu.

- Arrowheads and name labels on axes: and speaking of arrowheads, you can now put them on one or both axes. You can also attach the axis name label with a right-click of the **Context** menu.

- Reflex Angles: now we have them - for **Constraints**, **Annotations**, and **Calculations.**

- Point Trace - here's a fun new feature. Select some points and go to the **Context** menu to make them leave a trail when you drag them or animate the diagram. Check out the pantograph example!

- Edit the Color List - finally you can choose your favorite colors to use in the **Context** menu **Color List**.

- The Constraint Engine - has become more flexible making your drawings easier to define.

Where is the Math Illustrations Website?

Information on upgrades, additional technical support and loads of great examples can be found on the *Math Illustrations* website at: www.MathIllustrations.com.

Index

www.ingramcontent.com/pod-product-compliance
Lightning Source LLC
Chambersburg PA
CBHW051347200326
41521CB00014B/2503